The National Computing Centre

The National Computing Centre develops techniques and provides aids for the more effective use of computers. NCC is a non-profit distributing organisation backed by government and industry. The Centre

- co-operates with, and co-ordinates the work of, members and other organisations concerned with computers and their use

- provides information, advice and training

- supplies software packages

- promotes standards and codes of practice

Any interested company, organisation or individual can benefit from the work of the Centre by subscribing as a member. Throughout the country, facilities are provided for members to participate in working parties, study groups and discussions, and to influence NCC policy. A regular journal – 'NCC Interface' – keeps members informed of new developments and NCC activities. Special facilities are offered for courses, training material, publications and software packages.

For further details get in touch with the Centre at Oxford Road, Manchester M1 7ED (telephone 061-228 6333)

or at one of the following regional offices:

Belfast	1st Floor 117 Lisburn Road BT9 7BP	Glasgow	2nd Floor Anderston House Argyle Street G2 8LR
Telephone:	0232 665997	Telephone:	041-204 1101
Birmingham	2nd Floor Prudential Buildings Colmore Row B3 2PL	London	11 New Fetter Lane EC4A 1PU
Telephone:	021-236 6283	Telephone:	01-353 4875
Bristol	6th Floor Royal Exchange Building 41 Corn Street BS1 1HG		
Telephone:	0272 27077		

Introducing Systems Analysis and Design Volume 1

PUBLISHED BY NCC PUBLICATIONS

Keywords for information retrieval (drawn from
the NCC *Thesaurus of Computing Terms*): computer
projects, computer security, database, data control,
file organisation methods, forms design, systems
analysis, systems design, systems documentation

British Library Cataloguing in Publication Data

Lee, Barry
    Introducing systems analysis and design
    Vol. 1
    1. Systems analysis
    I. Title
    003   QA402
    ISBN 0-85012-206-6

First published in 1978 by:
NCC Publications, The National Computing Centre Limited,
Oxford Road, Manchester M1 7ED, England

ISBN 0-85012-206-6

This book is set in 10/11pt. Times Roman Series
and printed in England by Wright's (Sandbach) Ltd,
9 Middlewich Road, Sandbach, Cheshire, England.

Acknowledgements

The two volumes of Introducing Systems Analysis and Design derive from the combined efforts of several contributors. Thanks are due to George Penney, Norman Candeland, John Pritchard, Ray O'Connor (all of NCC), and Roger Jerram (then of NCC).

Particular thanks are also due to Barry Lee of Manchester Polytechnic, who was asked to organise and edit a large amount of information in a short time.

G L SIMONS
Chief Editor
NCC Publications

Foreword

Books on systems analysis usually abound with descriptions of computer hardware and software. For this reason, they often pose unnecessary difficulties for someone who is more concerned with improving the quality of information than with the application of the latest technology. This present volume is unique in that it confines itself to the non-computer aspects of systems analysis. These are the vital tasks which must, if the resulting system is to be effective and economic, precede the question of which computer to use, or whether to use a computer at all.

Any literate person should be capable of reading this book and of assimilating the vast amount of practical experience distilled here. This is particularly useful because of the undoubted need for a greater supply of literate people working in systems analysis: people who can write a succinct and convincing report, who can guide manager or clerk through the labyrinths of computer jargon to an appreciation of the real benefits that a computer, properly used, can bring.

I believe that this is the essential virtue of the book, with its particular orientation and style. I recommend it to all aspiring systems analysts, whether their background is in computing, in business, or in neither; to those already working in, or responsible for, the data processing function; and to line managers seeking to improve their function.

This book will help anyone to understand the purpose and practice of systems analysis. It will do so without detailed reference to the specialist jargon and proliferating devices and techniques, which only the computer expert needs to understand in depth.

GEORGE PENNEY
NCC Careers Projects Manager

August 1978

Contents

Part I
Orientation

Part I of the book aims to provide the reader with a background understanding of what systems analysis and design involves. It examines the role of the systems analyst and some approaches to the analysis and design of systems. It therefore acts as an introductory overview of the whole book and places the subsequent parts in their context.

The first chapter discusses some of the basic concepts of systems analysis and design by applying elementary systems theory to the organisation and by highlighting the function of information systems. Chapter 2 starts by examining the life cycle of an information system and then describes the role of the systems analyst in developing and implementing new information systems; particular emphasis is placed on the problems of achieving change in organisations. One of the most important factors in making changes to systems and procedures is effective communication of what is happening to all involved. Chapter 3, therefore, is devoted to discussion of the communication skills which the systems analyst must have to be successful in introducing new systems. The last chapter of this part is concerned with some of the high-level decisions that must be made before any new system can be developed. The issues involved are the choice of approach to system development, methods of involving users in the design of 'their' system, and the problems of assessing the feasibility of and justifying proposed systems.

1 The system in the organisation

SYSTEMS CONCEPTS

The starting point for a book on systems analysis and design must be a definition of 'system', a word used in many ways in everyday parlance. One speaks of an educational system, a computer system, the solar system, economic systems, and so on.

In systems analysis and design the concern is usually with man-made systems involving inputs, processes and outputs (the outputs being required by the system's objectives):

INPUTS ⟶ PROCESSES ⟶ OUTPUTS

For our purpose, a *system* can be regarded as:

a set of interacting elements responding to inputs to produce outputs.

The processes in a man-made system usually employ various types of element: physical, procedural, conceptual, social, etc. For example, a production system makes use of production equipment, production control procedures, 'laws' of production, and production employees.

The elements of the system define its boundary: the system is within the boundary; its environment is outside the boundary. It is often difficult to specify in detail the boundary of a system: the systems analyst defines the boundary to suit the purposes of the particular study. For example, one study of the production system may include raw material and finished products as within the boundary, whereas another study may exclude them.

15

Subsystems

Each system is composed of *subsystems* which are themselves made up of other subsystems; the subsystems, elements of the system to which they belong, are themselves defined by their boundaries. The interconnections between subsystems are known as interfaces. A subsystem at the lowest level, whose processes are not defined (the study of the system does not require it), is called a *blackbox* system; here the inputs and outputs are defined.

The systems concept requires the systems analyst to look at the system as a whole, but the full system may be too large to be analysed in detail. It is therefore necessary to divide or 'factor' it into subsystems, and then to integrate the subsystems.

Factoring into subsystems

The process of factoring continues until the subsystems are of appropriate size for the analysis/design project. The subsystems will usually factor into a hierarchical structure.

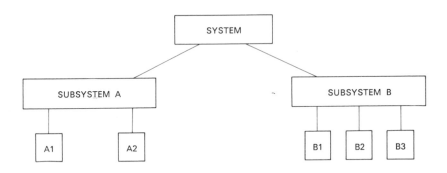

This hierarchy diagram does not show that each of the subsystems is an integral part of the subsystem or system to which it belongs. The problem of factoring is that it leads to a very large number of input/output interfaces between the subsystems. The number of interfaces is given by $\frac{1}{2}n(n-1)$, where n = the number of subsystems. Not all subsystems will interface, but clearly even a small number of subsystems will generate a large number of interfaces. One of the ways to overcome this problem is to group together those subsystems which have very close relationships. Such a grouping would be a likely design outcome in any case.

Integration of subsystems

The concept of integration is very important for the systems analyst: it draws attention to the primary importance of the whole system. It is the whole which dictates the role of the subsystems; the subsystems and their relationships should evolve from the concept of the whole, each being dependent on the whole system for its position and its relationship to other subsystems. The difficulty, as indicated earlier, is identification of the whole system. Theoretically, it is any system which behaves as if it is a single entity. In practice, the whole system is defined arbitrarily, often as a result of political or subjective decisions: in most organisations there is conflict over the objectives of the organisation as a system.

It should be clear that the major task of the systems analyst is to define the system in terms of its objectives, inputs, processes, outputs, boundary, and the interfaces between its subsystems.

The behaviour of systems

The second task for the systems analyst is to understand how the system behaves. A number of concepts assist in the analysis of behaviour.

Deterministic and probabilistic systems

A *deterministic* system is one in which the occurrence of all events is perfectly predictable. Given a description of the system state at a particular time, and of its operation, the next state can be perfectly predicted. An example of such a system is a numerically-controlled machine tool.

A *probabilistic* system is one in which the occurrence of events cannot be perfectly predicted. An example of such a system is a warehouse and its contents. Given a description of the contents at one time, and of the average demand, length of time to process orders, etc, the contents at the next point in time could not be perfectly predicted.

The systems analyst deals almost entirely with probabilistic systems.

Closed and open systems

A *closed* system is one which does not interact with its environment. Such systems are rare, but *relatively* closed systems are common. An example of a relatively closed system is a computer program which processes predefined inputs in a predefined way. A relatively closed system is one which controls its inputs, and so is protected from environmental disturbance. Closed and relatively closed systems are subject to increase in *entropy* or disorder, because they do not interact with their changing environment. For example, a firm which is not sensitive to customer demand will eventually decline.

An *open* system is one which does interact with its environment, being able to receive unexpected inputs. Open systems are organic in that by their nature they tend to react with the environment; a human being is an example of an open system. Open systems are usually *adaptive*, ie their interaction with the environment is such as to favour their continued existence. A firm which is sensitive to changes in customer demand, and in consequence adjusts its prices, changes its products or looks for new markets, is being adaptive. The systems analyst usually deals with adaptive, open systems and must aim to change them to make them more adaptive.

Stable systems

A stable system is one in which relationships are well defined but which, when disturbed by environmental factors, is capable of returning to its desired state. The stability is measured in terms of certain elements remaining within previously set limits. For example, a stock control system using a reorder level based on forecasts of demand would immediately cause a purchase order to be made when the stock reaches the reorder level. This is known as a *negative feedback mechanism* which records when the system is tending to deviate from its limits. Where the system has several such feedback loops (whereby, if one loop goes out of control another is activated to restore control), it is known as an ultra-stable system.

The systems analyst should set out to design systems which are stable. This may tempt him to design systems which are relatively closed and deterministic, because such systems are easier to design; but they are also inappropriate to man-made systems which should exploit the best features of humans and machines in achieving control.

Control in systems

The simple system above does not provide for effective measurement and control of the system. A suitable modification is shown:

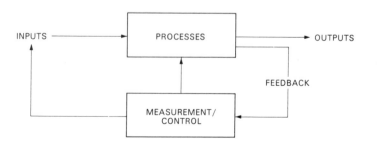

In this model, output is fed to the measurement/control function and compared to the desired limits. Any deviation from those limits causes a message to be sent to adjust either the processes or the inputs. (It is also quite possible for the control response to lead to a change in the system objectives or in the predefined limits.)

Control mechanisms, like systems, can be open or closed. An *open* control mechanism is one subject to unknown and unexpected changes, which would normally be associated with human control; a *closed* control mechanism (such as the thermostat on a central heating system) is automated and subject only to expected disturbances. Again, *open* and *closed* are the extremes of a wide spectrum.

The law of requisite variety

This law states that, in order to control each possible state of a system, there must be a corresponding control mechanism. For every way in which the system can get out of control, there must be a way of restoring control. This requirement may relate to a mass of information. For example, the stock controller in a retailing company with 80,000 stock-keeping units would need to have detailed knowledge of the state of each unit and be able to provide a control response for each state. This is an impossible task for one person: the work is usually divided among a number of subordinates, or a machine carries out some control activity.

The *law of requisite variety* suggests that one cannot use purely computer control for open systems. There must be a man-machine approach to the problem.

The principle of equifinality

According to this principle, an open system can reach the same final result from different initial conditions and by various routes. In other words a system can accomplish its objectives with varying inputs and varying processes; there is no one best way of achieving the objectives. The effect of introducing computer-based control systems may be to reduce the amount of equifinality. The systems analyst's aim is to design systems which promote flexibility of information utilisation in systems management.

SYSTEMS AND THE SYSTEMS ANALYST

A summary of the points made so far is now appropriate. The systems analyst is interested in systems; the systems dealt with will be man-made and capable of redesign. Systems will be analysed in terms of their objectives, and the inputs, processes and outputs required to achieve these objectives. The aims are to identify the boundaries of systems, their subsystems, and the interfaces between subsystems; and to factor into, and integrate, subsystems.

The systems analyst is also interested in the behaviour of systems, and will normally deal with human, probabilistic, open systems. If such systems are to adapt to their environment, the analyst must understand how control can be achieved. Within this overall framework, certain problem areas clearly confront the systems analyst:

– identifying the whole system and its boundaries;

– dealing with problems of interface;

– maintaining the system's dynamic interaction with its environment;

– building flexibility of control into systems;

– resolving conflicting objectives.

The usual environment within which the systems analyst confronts these problems is a business (which is taken to include both profit and non-profit organisations).

THE BUSINESS AS A SYSTEM

A business can be defined in systems terms. The system elements are physical (buildings, raw materials, finished products), procedural (order processing routines, credit checking procedures), conceptual (statement of policy, market for products), and social (workers, departments). The system operates in an environment and relates to that environment. The environment includes various elements: physical (transport routes, shops), procedural (codes of practice, legal requirements), conceptual (monetary system, political ideologies), and social (trade unions, suppliers, customers).

The business can be factored into a number of subsystems, normally the functional areas. These include those dealing closely with the environment (the marketing and purchasing subsystems), those dealing with transfor-mation functions (the production subsystem), and those acting in a supportive role (the accounting, personnel, management services sub-systems). These subsystems are all interconnected, designed to reflect the overall organisational objectives. A simple model of the subsystems of a business showing some of the relationships is shown (fig. 1.1). The interfaces within the business are indicated, with illustration of the difficulty in defining boundaries: should purchasing be a subsystem of production?, should distribution be a subsystem of marketing rather than production?, etc.

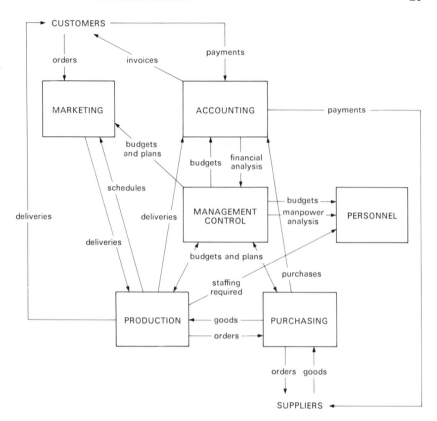

Figure 1.1 Simple Model of Business Subsystems

The functional areas of the organisation will be determined by the nature of the organisation's business. In general the areas are characterised by the need to satisfy the demands of a particular group of customers or clients. These demands can be satisfied in a number of possible ways, eg:

- manufacturing and selling a product which has been produced from raw material or by further processing on products purchased from other organisations;

- purchasing and distributing finished products (wholesaling and retailing);

- offering a service (eg, hotel and catering, education, banking, insurance, cleaning, maintenance activities).

Such activities are carried out by different types of organisations and the nature of the organisation will also affect the organisational subsystems.

Private enterprise firms operate within a competitive economy (sometimes constrained by cartel and monopoly). While a main purpose is to satisfy the producer/consumer relationship, it is necessary to survive financially and provide a minimum return on invested capital. Operations are characterised by private capital investment (and government aid), the profit motive, and investment risk (more relevant in small firms than in the multinationals).

State-owned businesses (steel, coal, electricity, rail, etc) are characterised by state capital investment, a virtual monopoly of the home market, and state intervention. These businesses may still have to compete for trade in world markets, and for people and capital resources at home. Their systems and subsystems still need to be controlled similarly to those of private enterprise: performance is controlled by accounting systems and measured by the profit-and-loss concept.

There are also various non-commercial businesses such as charitable organisations, foundations, educational establishments, and government departments (health, water supply, police, fire services, etc) financed by, and accountable to, central government or local authority.

Objectives

Particular business objectives derive from the general reasons for company existence. In the case of joint stock companies the objectives are broadly defined in the *objects clause* of their memorandum of association. The clause states every type of activity which the business intends to undertake, the trading area of the business being broadly defined.

It is the responsibility of corporate management to define specific objectives: commodities or services to be produced and distributed; the methods of achieving the objectives – all within the financial constraints of solvency and expected return on capital investment.

The procedures designed to achieve the objectives become the business systems and subsystems, usually related to corporate long-term plans. Main objectives are usually static, but subsystems are dynamic, changing as more effective techniques become available. Business systems are subject to internal and external influences, such as new legislation, trade fluctuations, company mergers, management decisions, market developments, etc.

Controls

The importance of effective planning is self-evident. It is essential to check how effectively the plans are being carried out and the objectives being achieved. Control systems for this purpose operate at many levels and are characterised by:

- setting of plans and targets;
- measurement of performance;
- comparison of performance with plans;
- adjustment of plans as necessary.

The overall process, facilitating the review and adjustment of plans and targets, is continuous. It is important that control systems are effective in rectifying deviations from plans. Otherwise the controls simply waste money and introduce delays.

In any organisation several control systems will be in operation ranging from overall management control information to specific systems for stock control, budgetary control, production control, credit control, etc. In addition, the accounts of an organisation are subject to audit control by outside auditors.

Organisation structure

In order to achieve its objectives, the organisation must be structured on principles having the benefits of specialisation without losing overall co-ordination and integration; and the formality of the structure must be tempered by the need to allow (and encourage) informal contacts between staff.

The traditional principles on which organisations have been structured aim to achieve parity of authority and responsibility; harmony of objectives (all staff working to a common end); unity of command (each person having only one immediate superior); unity of direction (one manager and one plan for each major objective or set of objectives); appropriate span of control (the number of subordinates which one person can manage); and control by exception (feedback of deviations from an agreed plan).

In recent years increasing attention has been devoted to the *informal* structure of the organisation, ie the relationships which people make which cut across the formal structure. Often the informal structure reflects more closely the power structure of the organisation, and certainly it facilitates quicker and smoother operation at all levels than does the formal structure. The *formal* structure defines nominal responsibilities and authority in relation to the use of resources. A typical organisation structure for an engineering company is shown (fig. 1.2).

The formal organisation structure, tending to be rather static, is important to the systems analyst: information flows are often determined by the organisation structure.

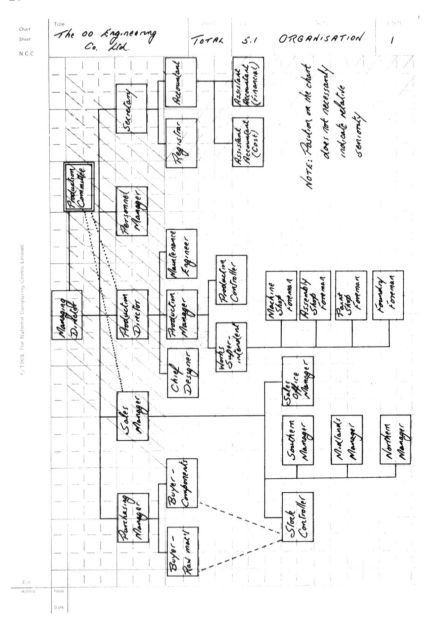

Figure 1.2   Organisation Structure of an Engineering Company

The systems analyst needs to be aware of relationships between people and to realise that boundaries of departments are usually jealously guarded. Various approaches to organisation structure have been adopted:

Line organisation   This is the simplest form of business organisation: the lines of communication flow directly through a hierarchy from top management downwards (fig. 1.3).

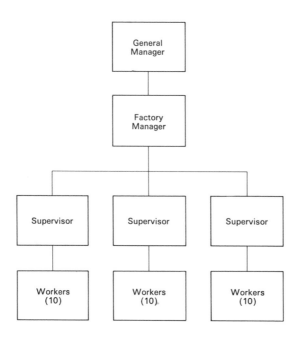

Figure 1.3   Line Organisation

Functional organisation   Functional (or staff) organisation is usually adopted where business efficiency can best be achieved by 'one person' management responsibility being transferred to more than one person, according to assigned supervisory specialisation. Authority still flows from the top but supervisory staff, organised functionally, have a degree of authority over all personnel at the level below. The aim is to achieve efficiency by promoting co-operation between specialised supervisors (fig. 1.4).

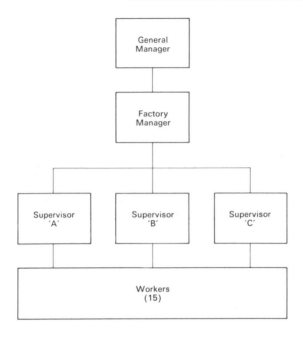

Figure 1.4   Functional Organisation

Line and staff organisation   This approach characterises most large businesses. Here the range of management skills required are too diverse to be found together in individual line managers. Specialist staff are expected to define strategy, propose solutions and prepare plans for line management. The features of both line and functional structures are combined. Delegation of authority in a direct line ensures control; and staff specialisation provides accountability through special responsibilities (fig. 1.5).

A more modern approach, as yet not widely applied, is 'matrix organisation' which allows staff to work in product groupings which achieve lateral relationships across the organisation and yet to have line responsibility within a functional area.

There has been a wealth of research into organisation structuring which this book cannot begin to examine. The systems analyst is urged, however, to become familiar with current thinking on management and organisation theory.

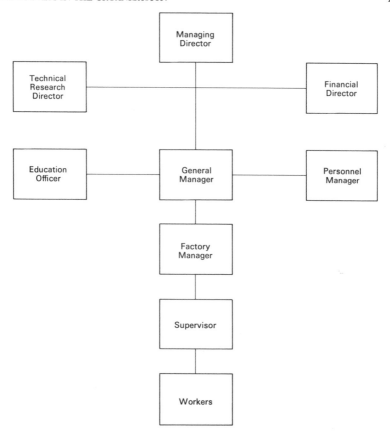

Figure 1.5   Line and Staff Organisation

Hierarchical structures

A problem of hierarchical structures within organisations is that they discourage the flow of consistent and accurate information to the top of the hierarchy. Different departments, making use of the same basic information, adopt different approaches to problems. Thus, reports which are summarised from a number of departments as they move up the hierarchy may be based on different timescales, different interpretations of rules, different coding systems, etc.

Departmentation encourages empire-building, which in turn leads to duplication of effort and information; for example, lack of trust (or inappropriate organisation structure) may cause the sales department, the accounts department and the production department each to maintain details of customers' orders. The production of *ad hoc* reports (to meet a sudden specific enquiry from top management) which cross departmental boundaries is often frustrated.

Hierarchical structures tend to encourage formal, highly structural information flows in which information is inconsistent, incomplete, and filtered (possibly in the wrong way).

INFORMATION SYSTEMS

The subsystems of the business have been considered as related to specific business functions, such as marketing, production, and accounting. The business has been described as a system which receives resources (capital, people, plant and materials) and produces goods or services.

In order to work as an effective unit the business has to make use of information: an information system, often regarded as another subsystem, is superimposed over and implanted in all the subsystems. The information system is concerned with procedures for the storage, control and flow of information which passes between the functional subsystems to ensure that common and accepted approaches are taken to the tasks which are presented to the organisation.

The information system provides information for decisions and control, and acts as linking mechanism between the functional subsystems. Since the advent of computers, information provision (known as data processing) has become an organisational function in its own right.

Data processing

Data denotes facts of the business represented by numbers, alphabetic characters or symbols signifying condition, value or state. Data may be an amount on a bill, a customer's name or reference code, an address, or a stock item reference. Data is processed (ie analysed, summarised or assembled meaningfully) to provide *information*.

Data processing refers to this conversion of data to information. Electronic data processing (EDP) refers specifically to computerised data processing systems.

The computer can use the same data which has been input or stored for the payroll as the basis for producing a management report (on, say, manpower forecasting). This is the concept of integrated data processing. An operation such as sales invoicing is integrally linked with order processing, delivery documentation, sales forecasting, inventory control, sales statistics, and the sales ledger. This emphasises the importance of integrated systems planning based on the subdivision of the total system into meaningful subsystems.

Use of a computer in this way has two implications. Firstly, as the system is developed it affects more and more parts of the business. Secondly, it is only top executives of the company who can determine the effects of computer development on the business. For example, only they can determine in an inventory control subsystem whether control of bought-out stocks should lie in the hands of production or purchasing departments.

Computer costs and processing capability have traditionally led to centralisation of the data processing function. More recently the decreasing cost/power ratio of hardware, and the greater understanding and acceptance of computers by line management, are stimulating a trend towards decentralisation of hardware: it is now feasible for each function of the business to have its own computer.

Because of the developments in technology, and because of the enormous cost of developing a total information system for the organisation, most businesses have developed a number of information systems rather than one integrated system. Examples of such systems are usually functionally oriented (production information systems, sales information systems) or even problem-oriented (sales order processing systems, stock control systems).

The development of any information system affects functional areas beyond the one directly involved. Emphasis has therefore been placed on achieving integration of data held on files, so as to co-ordinate all information systems. Figure 1.6 illustrates this principle.

The database is seen as the store of data needed by the various subsystems of the organisation. The database may be centralised (ie held in a common pool) or decentralised (ie distributed to separate pools associated with individual subsystems), but its design will take place in a co-ordinated way. The database serves the transaction processing activities of the operating systems (ie, payslip production, order processing, sales ledger), the management control activities (ie reports of stock levels, sales, scrap, cash flow), and the corporate planning activities of top management. The data of the organisation is available in a consistent and accurate form to all levels. These data processing considerations are discussed in subsequent chapters.

A production planning information system

To give some idea of the role of the information system in a business organisation, it is perhaps best to look at a typical system, eg production planning. In an organisation where sales are made from stock, the role of

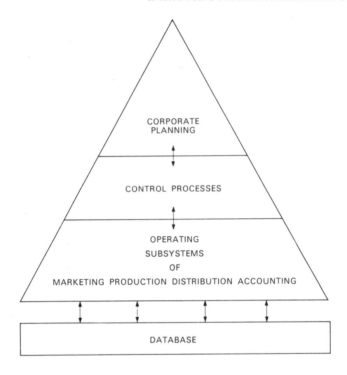

Figure 1.6   The Business System

production planning is to ensure that sufficient stock is available. A simplified chart of the information flows might look like this:

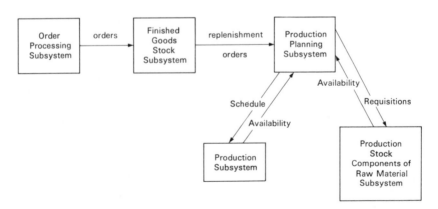

Thus an information system for production planning must have available to it as input, data on customer orders, production availability, production stock availability: this data is of course required or generated by other subsystems. (The data is also valuable for company planning and control purposes.) The system processes this data by expanding the orders into material requirements, determining the production processes required, and working out a production schedule.

It produces, as outputs, material requisitions to the production stock subsystem and a production schedule to the production system. Both of these are outputs which can change from day-to-day in the light of priorities, order volumes, machine availability, manpower shortages, etc, and so the system must be able to reschedule at regular intervals.

Clearly the information system needs to know a lot about orders, forecasted demand, methods of production, resources required, timing of processes, etc. It is a dynamic and open system requiring human involvement. Above all, it needs to be fully integrated with the systems with which it interfaces.

SUMMARY

Systems analysis is concerned with investigating, analysing, designing, implementing and evaluating information systems in organisations. In order to do this, the systems analyst needs a framework within which to examine the operation of organisations; and the concepts of systems theory provide a suitable framework. Organisations are dynamic, complex, open, man-made systems whose participants have conflicting objectives. The information system of the organisation should help to achieve integration of its subsystems and to overcome the problems of hierarchical organisation structures. The systems analyst must design information systems which meet organisational objectives, promote integration of activities, facilitate control, and which are flexible and robust. The use of computers in processing data has greatly enhanced the opportunities for building such information systems, particularly by allowing central access to stored data. This chapter presents some of the underlying concepts which the systems analyst encounters.

2   The systems analyst

INTRODUCTION

The systems analyst is concerned with analysing systems with a view to making them more effective either by modification or by substantial redesign. The systems involved will invariably be business systems of the kind described in the last chapter, and usually the computer will be under consideration as an aid for improving the system's operation. The term 'systems analysis' was used, long before computers appeared, to describe such activities as operational research (OR), organisation and methods (O & M) and work study (WS).

With the advent of the computer as a data processing tool, systems analysis has been increasingly applied to the techniques of determining how best to apply the tool, ie analysing the existing situation of an organisation, in terms of a system and its subsystems.

The range of activities that comes within the concept of systems analysis is not always clearly defined. Some organisations split the functions of systems analysis and systems design, by recruiting and training as *information analysts* their own divisional staff. They are trained to record and analyse information on existing systems, and to prepare specifications of requirements for proposed systems. These proposals and specifications are then passed either to *system designers* in the data processing department, who are technically skilled in designing computer-based systems, or to an independent computer bureau.

In other organisations, systems analysts may be responsible for system analysis, outline system specification, and the detailed design of the computer system. There are many variations on this theme. In the present book the term *systems analysis* relates to all the functions of analysis and design.

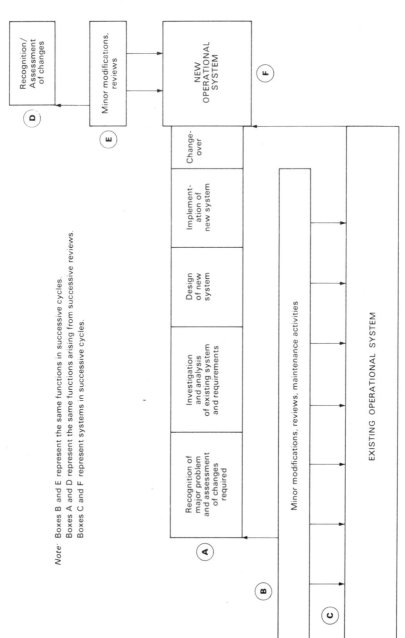

Figure 2.1 Life-cycle of Systems

SYSTEMS LIFE-CYCLE

It is in the nature of systems that they share a common life-cycle pattern. After a system has been in operation for a number of years, it gradually decays and becomes less and less effective because of the changing environment to which it has to adapt. For a time it is possible to overcome problems by amendments and minor modifications to the system but eventually it will be necessary to acknowledge the need for fundamental changes. At this stage the systems analyst becomes involved and investigates the problems and the requirements for a new system; once the requirements have been adequately identified, a new system can be designed and subsequently implemented.

The new system will operate for a number of years until it too becomes obsolescent and the cycle begins again. Figure 2.1 illustrates this cycle of activity. At any stage in the cycle, there will be an operational system which will be subject to continual modification, review and maintenance (either by the system operators or by specialist staff). At some point in time a review of the system will indicate the need for specialist investigation of the problems, especially if it is envisaged that a computer may assist in solving the problems, and a systems analyst will be called in to begin the development of a new system.

System development activities

A more detailed identification of the system development stages is shown in figure 2.2.

Establishment of terms of reference

The first stage of any project, sometimes called the Preliminary Assessment, is a brief investigation of the system under consideration to provide, to the organisation's computer steering committee and any project team, a set of terms of reference for more detailed work. It will be carried out by a senior manager and will result in a Study Proposal.

Initial study

If the Study Proposal is accepted by management, it will lead to an investigation of the existing system or problem area, conducted in close collaboration with user management and in sufficient depth to establish in broad terms the technical, operational and economic feasibility of proposals. It should also define the resources needed to complete the detailed investigation which would follow acceptance by management of the initial findings.

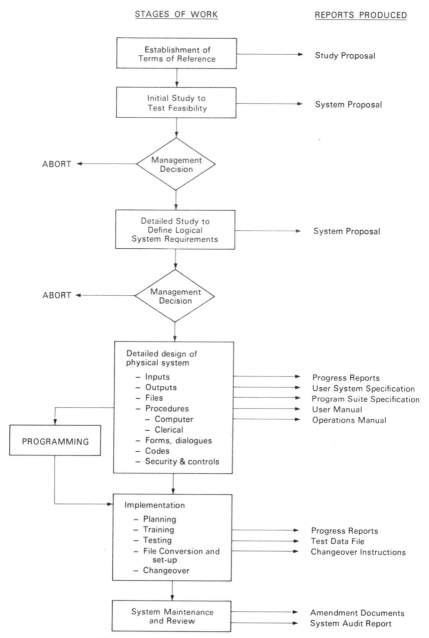

Figure 2.2   System Development Stages

Detailed study

Detailed investigation should be conducted with personnel closely involved with the area under investigation, according to the precise terms of reference arising out of the initial study report. The tasks to be carried out should be clearly defined, eg:

- examine and document the relevant aspects of the existing system, its failings and problems;

- analyse the findings and record the results;

- define and document in outline a proposed system;

- test the proposed design against the known facts;

- produce a detailed report to support the proposals;

- estimate the resources required to design and implement the system.

The objectives at this stage are to provide solutions to stated problems, usually in the form of a specification to meet the users' requirements; and to make recommendations for a new computer-based system.

Analysis is an iterative and progressive process, examining information flows and evaluating various alternative design solutions (under the headings of operational, technical and economic feasibility) until a preferred solution emerges. It will be documented as a system proposal.

Detailed design

System design is a creative as well as a technical activity, including the following tasks:

- appraising the terms of reference;

- appraising the analysis of the existing system, particularly regarding problem areas;

- defining precisely the required system output;

- determining data required to produce the output;

- deciding the medium and format of files;

- devising processing methods and use of software to handle files and to produce output;

- determining methods of data capture and data input;

- designing forms;

- defining detailed clerical procedures;

– calculating timings of processing and data movements;

– documenting all aspects of design.

The system should be designed to cope with all likely errors, and any subsequent changes and modifications. When the user system specification is complete, it should be presented to management with a proposal which outlines the main features of the design, and provides a restatement or revision of the objectives, costs, and benefits to be expected. Management should then formally accept or revise the proposal before further development. When agreement is finally reached, a program suite specification can then be handed over to the programmers for conversion of the computer parts of the system into programs of coded instructions appropriate to the computer. Meanwhile, the systems analyst will produce detailed User and Operations Manuals for the system.

Implementation

Implementation is concerned with those tasks leading immediately to a fully operational system. It involves programmers, users and operations management, but its planning and timing is a prime function of systems analysis. It includes the final testing of the complete system to user satisfaction, and supervision of the initial operation of the new system.

Maintenance and review

Once the system has settled down and been running for some time, the systems analyst may be involved in amendment procedures to adapt the system to changing conditions; in auditing the system to check that the stated objectives of the system are still valid in the present environment; and in evaluating the achievement of those objectives.

This list of stages is not definitive. Each organisation adopts its own approach to system development and the approach will be defined in its data processing standards. For example, some data processing departments will merge the initial and detailed studies together into an investigation and outline design phase which includes an assessment of feasibility; others will not bother about feasibility in certain circumstances. In any case, it should be realised that the whole process of system development is iterative; investigation, analysis and design activities are carried out continually. During investigation, ideas for improvements and redesign will emerge; and during detailed design, a return to the user department will often be necessary.

The detailed activities involved in each of the stages are described in the remainder of the book (vol. 1 and vol. 2). The project reports which are shown by name (fig. 2.2) are discussed in detail in volume 2, chapter 22.

THE DATA PROCESSING DEPARTMENT

The system development activities described above, and the operation of the computer part of any computer-based system will be carried out by staff in the Data Processing Department.

The data processing function can be organised in a variety of ways. Conventionally, the dp department is headed by a data processing manager (DPM) who is responsible to senior management at policy making level. The DPM decides policy within the department to ensure that the data processing functions are all working towards the same end. These functions include:

- management;

- systems analysis and design;

- computer programming;

- computer operations;

- data preparation;

- data control.

A typical organisation chart for a data processing department is shown (fig. 2.3).

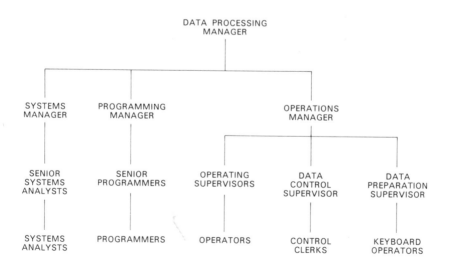

Figure 2.3   Data Processing Department Organisation Chart

Systems Analysis

The systems analyst (or systems designer) designs and implements systems and procedures throughout the organisation.

Programming

Programmers convert written specifications, prepared by systems analysts, into coded instructions in the format required by the computer. Available computing facilities and techniques are exploited to produce effective programs properly tested and documented. Subsidiary and sometimes extensive, activities are concerned with amending existing programs to meet changing requirements. Completed programs are passed to Computer Operations for running on the system.

Computer Operations

Computer operations staff are responsible for the day-to-day computer processing of operational systems. This comprises:

- computer operating;

- data control;

- data preparation.

Computer operating includes operating the computer, its peripherals, and other ancillary equipment. This requires a working knowledge of the hardware, the operating system that controls the basic computer operations, and the operating instructions for each program being processed. The operators must know the action to be taken in the event of malfunctioning of the equipment or when error conditions arise from the program being used.

Data control supervises the passage of jobs through the department in accordance with working schedules. This comprises:

- receiving raw data from other departments and controlling the movement of data into and out of data preparation;

- checking the movement of data into and out of Computer Operations, making adjustments for rejected data, and reconciling with controls;

- ensuring that computer output is despatched to the correct addressees.

Data preparation converts raw data to the appropriate computer input medium for subsequent processing, ie punched cards, paper tape, magnetic tape or disk. Similarly, written program instructions are converted to the appropriate input medium.

The organisation chart for a large Computer Operations Department is shown (fig. 2.4).

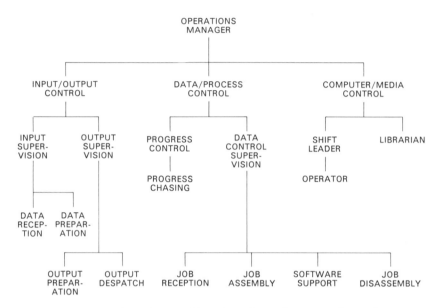

Figure 2.4 Organisation Chart for a large Computer Operations Department

Project teams

The design staff involved in the design of computer-based systems are usually organised into project teams. While individual team members are responsible to their particular managers for administrative purposes, as team members they are individually responsible to a project leader for the design of a specific system. There are various types of project organisation (figures 2.5 and 2.6 illustrate two possibilities).

Although a project leader is normally a senior systems analyst, a manager from the user department is sometimes appointed as the leader. Alternatively, user representatives may be included in the team.

Management Services Department

As the role of the data processing department has grown in importance, there has been a tendency to locate the data processing function within a management services department. Such a department would have re-

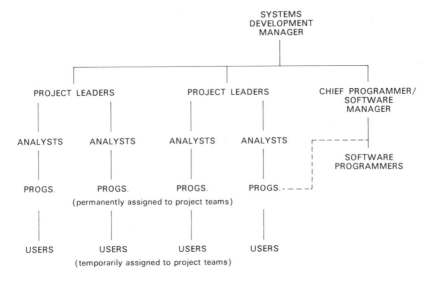

Figure 2.5   Project Organisation (1)

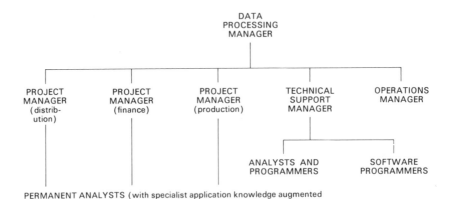

Figure 2.6   Project Organisation (2)

sponsibility for all information and planning activities within the organisation. These may include organisation and methods, operational research,
and corporate planning staff. A special post may also be created for

training and standards, and another for database administration. A possible structure for such a department is shown (fig. 2.7).

The systems analyst who operates in this type of structure, will have to liaise with, and take advantage of, the skills of the specialists in O & M, OR, database, and training. Thus several of the activities which are covered in this book, as belonging to the systems analyst, will be carried out by specialist staff. In relatively small organisations, the reverse will be true: the analyst may carry out more functions, including programming.

Another development which has taken place in some organisations is to separate computer operations from systems development. This has various advantages: it ensures that formal procedures are adopted for passing work from one department to the other; it facilitates equitable control of costing and pricing of services supplied; and it enables more objective assessment of priorities both in operations (between development work and operational work) and in development (between existing computer-based systems and new ones).

Whatever the structure adopted, it is clear that the data processing function as a whole should not report to a line manager. The development and operation of computer-based systems should be achieved in the light of overall organisation (system) objectives. These can only be assessed at corporate level.

THE ROLE OF THE SYSTEMS ANALYST

It is probably true to say that systems analysts work on the frontiers of knowledge, not in the true scientific sense but insofar as their own organisations are concerned. They lead their organisations into new ways of thinking: their job is essentially creative. They are likely to assume a variety of functions, many of which require management skills; and they are also specialists in a new and expanding discipline.

The actual tasks to be carried out depend on the terms of reference; nevertheless, systems analysts are likely to be called upon to perform any of the tasks relevant to a particular project. They are also likely to meet problems which are outside their experience, especially in user departments. They must work in a consultative and advisory capacity within user departments: they cannot assume any executive authority within those departments.

An agent of change

The systems analyst is an agent of change; yet the system environment is resistant, even hostile, to change. Resistance will usually exist from the start of a project, firmly fixed in user departments. Systems analysts can only be

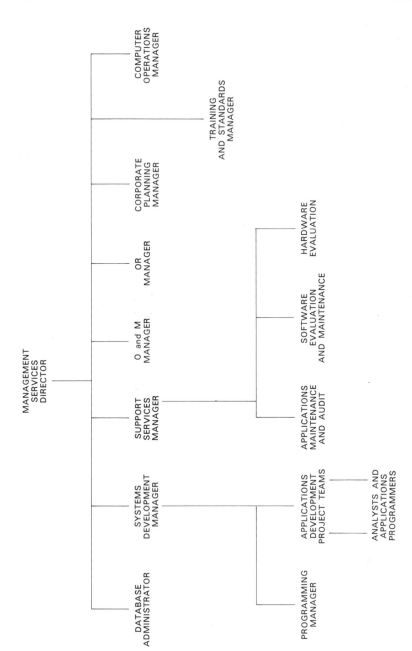

Figure 2.7 Management Services Department Organisation Chart

as effective as the amount of change they can achieve. Their reports may contain good ideas, but unless these are accepted and implemented with goodwill by user staff, little will be accomplished.

The objective of systems analysis is not merely to generate new ideas and procedures through an intellectual technical approach; it is also a matter of gaining acceptance of these by the people who will put them into practice. The analyst needs to be aware of the importance of human relations, especially when presenting and implementing new ideas that are likely to change people's routine.

Users are often committed to existing systems. Having designed and maintained them with long term stability in mind, users will naturally be apprehensive about changes based on unfamiliar technology.

Managers are being expected to come to terms with a technology which was unknown when some of them were trained. In addition, computer-based systems have not always had a good press, failures often being publicised more than successes. And there can be justification for critical attitudes: studies suggest that more than 50 per cent of computer installations are unsuccessful, costs not being recovered and expected benefits not being achieved.

The impact of systems analysis can be extremely disturbing to a manager. Systems analysts, often computer oriented, do not always give full consideration to the needs of users, management or the organisation. Resistance on the part of management may have value. Managers have experience vital for the effective operating of a new system. It is not unknown for a systems analyst to spend an afternoon discussing a manager's requirements and then to go away assuming that the distilled knowledge of many years of experience has been acquired. Managers naturally resent this attitude.

Systems analysis, when carried out with intelligence and understanding, can be rewarding to both the analysts and the organisation. To operate effectively, systems analysts must win the confidence and continued support of management throughout the development and subsequent operation of new systems. They should establish a reputation for integrity, impartiality and helpfulness: their approach should be planned to avoid antagonism.

Human aspects of change

People have individual systems of belief based upon experience, education, emotion, intelligence, knowledge and interests. Intelligent logically-minded people may have radically different views on the same subject. Different attitudes have developed through the years and are based partly on assumptions about the non-measurable factors in their environment.

People tend to believe what they want to believe. Some common illogicalities are: generalising from particular cases, assuming cause and effect from mere correlation, attributing a logical reason to an emotional belief, and transferring subconscious dislikes to people or things that serve thereby as scapegoats.

The main motivating forces in mankind are generated by basic human objectives, considering all possible objections and the alternatives that can be offered. Timing should be such that the degree of persuasion will be effective. They must plan how to get users to criticise their present situation and create a desire for change. The following suggestions may be useful:

- do not expect to convince people instantly;

- encourage participation in decision-making; this encourages commitment to decisions;

- emphasise essentials and avoid using too many arguments;

- ask questions that emphasise areas of agreement;

- aim at a mutually satisfactory solution rather than a total conversion;

- avoid criticism of the past; concentrate on positive aspects of the change and the common desire to make progress;

- listen sympathetically to problems and objections, but do not assume that verbal objections are necessarily the real ones; they may be rationalisations of emotional objections that the person knows are irrational and will therefore not admit;

- be wary of using negative suggestion, such as, 'it would be fine if we could do so and so, but it is not really possible in a firm like ours'. This approach may encourage an aggressively favourable reaction to the implied challenge, but if it fails it creates a situation from which it is difficult to recover.

Once agreement seems reasonably certain, get the plans, including a programme and timetable, accepted quickly. Emphasise continued support and help, both during and after the change. Give management and supervision credit for the change, and remember that to be logically right can sometimes be psychologically wrong.

Sometimes people resist persuasion by:

- not listening;

- attributing ulterior motives to the persuader, preferably behind one's back to others who may be affected;

- concentrating on disliking the persuader;

- exaggerating objections, especially the danger of repercussions and the unsuitability of timing;
- raising the temperature of the discussion and discussing personalities wherever possible;
- keeping real objections to themselves;
- maintaining prejudices.

People may resist persuasion for various reasons:

- fear: of losing their job, of wage reduction, of inability to learn a new job, of loss of prestige, or simply of the unknown;
- suspicion of management's motives in making the change;
- resentment against what appears as a personal attack, or a feeling that any change is a personal criticism of the way the job is being done;
- social upset caused by dissolving a working group.

Ways of overcoming these background reasons for resistance include:

- keeping people informed well in advance, indicating the full reasons and the benefits;
- giving people the opportunity to participate by making suggestions;
- giving security, which may mean obtaining a guarantee from management on the financial future, or a promise of retraining;
- creating a favourable atmosphere; giving people time to get accustomed and trained for change before implementing it (there are some rare occasions where the quick introduction of change without notice may be best).

Personal characteristics

Having outlined the character of systems analysis, and considered its main functions, it is worth looking at the personal qualities a systems analyst needs in order to succeed.

Systems analysts need the capacity to assimilate the complexities of systems and to reduce them to their fundamental logic. It is important to have insight and to avoid hasty conclusions.

The development of a new system involves the utilisation of expensive resources, and its results may have a wide-ranging effect on other systems and their resources. Systems analysts must therefore produce sound plans, appreciating the relevance of new information. This demands an orderly and disciplined approach, particularly as there may be no direct supervision.

Plans of action have to be maintained despite the inevitable setbacks: the analyst requires persistence in overcoming difficulties, stamina, strength of character, a broad sense of purpose and a flexible outlook.

It is also required that thoughts, ideas and proposals be clearly expressed in both speech and writing. Analysts need to be relaxed listeners. To maintain control through numerous interviews and meetings, they must be accurate and precise conversationalists with more-than-average social skill in communicating and working with others. In addition, as they will spend some time discussing and presenting proposals to user management, they must be able to sell their ideas effectively, appearing confident and enthusiastic.

Career recruitment

Aspiring candidates to a career in systems analysis may be found:

- already employed in one of the other branches associated with data processing, eg, O & M, OR, programming;

- in operating departments of commerce and industry, eg accounting, engineering, planning;

- as graduates, especially those with business or computer-related degrees.

Whatever their background additional qualifications and experience will be needed. Suitable candidates may have already acquired the necessary computer techniques. They must also acquire the ability to communicate at all levels using appropriate media, and gain familiarity with the business areas of the organisation.

The experience of candidates from a business background will vary: from a specialised knowledge of a restricted area (such as production control in the steel industry) to a wider overall understanding of business (such as management information systems). Business experience may contribute maturity and a broad base when combined with the other skills, particularly in computer technology.

Computer technology is of course a basic requirement in the complete systems analyst. Training is needed in the use of computer hardware and associated peripheral equipment, and of software. An appreciation of available applications packages is also necessary. The necessary communication with programmers demands an understanding of programming principles.

A full list of relevant areas of knowledge might include:

- data processing techniques;

- organisation and methods;

- information and communication theory;

- communication skills;

- business studies and accountancy;

- organisation theory;

- operations research;

- quantitative methods;

- industrial psychology and sociology;

- ergonomics;

- project management;

- systems concepts.

Essential abilities

The personal characteristics and areas of knowledge described above are background needs for the systems analyst. To operate skilfully in the job, certain essential abilities are required. The first of these is the ability to recognise, identify and define problems. Usually the analyst works in an area of complex uncertainty in which managers feel that they have a problem (or a need for more information) but cannot pinpoint it.

The analyst has to be able to assist the manager by a logical, objective, fresh analysis to clarify the requirement. Secondly, the systems analyst needs to be able to develop a number of possible solutions to the problem which meet the technical, economic and social demands and to evaluate them. There is never a 'right' answer, and so the analyst has to use judgement, and knowledge of a wide variety of disciplines to produce the best solution in the particular circumstances. Thirdly, the systems analyst has to be able to carry out this intellectually demanding, theoretical and creative activity within real-life situations, in which there are pressures of time, money and people. It is necessary to accept and attempt to reconcile many different views of the world, and to produce a workable system within time and cost constraints.

SUMMARY

Artificial systems have to be designed before they can operate (though sometimes the design evolves from the operation), and have to be redesigned or modified when their operation ceases to be effective or efficient. The systems analyst is involved mainly in developing new information systems; this activity involves investigation of current systems, proposing and evaluating possible new systems, designing in detail the new system which is agreed, implementing it, and maintaining it during its operational life. The systems analyst is one of a number of people involved in information system activity and is usually located in the data processing department of an organisation. The major role of the systems analyst is as a catalyst, encouraging information system users to be critical about their current information systems, to identify their information requirements, and to define a new information system to meet their needs. Because of this, and because system development inevitably involves change, the systems analyst needs to be a very good communicator.

3 Communication skills

INTRODUCTION

Much of the system analyst's 'stock-in-trade' is concerned with the creation and expression of ideas which have to be imparted to others.

The systems analyst has to communicate with other analysts in project teams, with programmers, with clerical and shop-floor staff, with user managers, with computer operations staff, and with people outside the organisation, such as auditors. The analyst has to be able to 'sell' ideas to people, convincing them that the new system will be an improvement worth the risk, and cost-beneficial. It is necessary to present ideas clearly so that the changes which a new system will cause will not result in chaos. The analyst has to keep people informed of developments, maintaining their interest and commitment. In all of these activities, effective communication is not merely the *passing on* of information, but its *presentation* in a way that will be comprehensible to the recipients and likely to gain acceptance for proposals.

The basic methods for communicating to others are written reports, presentations and meetings; these may be required at all stages of a project. They are of special importance when education or training activities are taking place. Aids to support these methods are standard documents and techniques. This chapter examines methods and aids, looking first at some communication problems.

PROBLEMS OF COMMUNICATION

The systems analyst needs to realise that communication is one of the most difficult areas in which to achieve effective performance: there are complex variables which affect the communication process. For example, at each stage of a simple communication between two individuals several things can go wrong.

The originator of the message has to couch it in a language which both parties can understand; this may cause semantic problems. The message is then emitted via some physical medium (voice, hand, telephone, etc). This may cause physical distortion of the message. As the message is being transmitted, external interference may prevent it from being heard or read accurately. At the time of receiving the message, the recipient may be distracted by some other event. Furthermore the recipient may consciously or otherwise, for a variety of social and psychological reasons, filter or block out certain elements of the message. Interpretation of the message received may differ considerably from that which the originator intended.

The systems analyst needs to be aware of the various barriers to successful communication of ideas in organisations. These include:

- organisation structure, personnel status and official procedures which can prevent messages from reaching the appropriate recipient;

- interpersonal barriers, such as values and attitudes, which can cause a gulf between the people involved in the communication process;

- individual barriers created by personal habits and idiosyncrasies;

- barriers created by differences of time, location and technology;

- overload, which occurs when people have too much information to handle and start to filter or block messages sent to them.

Careful selection of the channel of communication is of crucial importance, and the analyst needs to distinguish between *formal* channels (which follow the hierarchical structure, are official and recognised, tend to be well organised and standardised, and convey authority) and *informal* channels (which tend to have a socio-emotional function in achieving more personal contact on a spontaneous and unstructured basis). The channel should be chosen to achieve reliable communication, involvement on the part of the recipients, ease of communication and retention of the message. Above all, the communication channel must be two-way to allow feedback; if no feedback occurs, then communication has not taken place. Often, different channels can be used to complement one another. The formal report can be backed up by a presentation and discussion of its contents.

To overcome some of these problems the systems analyst needs to determine the answers to the following questions:

- who is the originator? What is the originator's background, status, knowledge, jargon, attitudes, interests, etc?

- what is the message? What is the content and how can it be expressed?

- who is the recipient? What is the recipient's background, status, knowledge, jargon, attitudes, interests, etc?

- what is the purpose of the message? For what change in the recipient is the originator aiming?

- in what circumstances will communication take place? Can the circumstances be changed to facilitate communication?

- what method of communication will be used? Report, formal presentation, informal discussion, letter, notice, telephone call, etc? What problems will the defined method create and how can they be resolved?

This list of questions will force the originator to be more sympathetic to the needs of the recipient and make the communication more effective. One last observation needs to be made: the effective communicator learns from past experiences, and each report or presentation should subsequently be reviewed for its effectiveness.

WRITTEN REPORTS

There are two basic kinds of written and formal communication with which systems analysts will be directly concerned:

- *reports*, written to persuade management to authorise proposals for change;

- *reference manuals*, to provide users and computer operations staff with a description of a new system, together with instructions to cover all likely events.

This section is concerned with principles of report and manual writing; detailed contents of appropriate reports and manuals are covered in Volume 2, Chapter 22.

Reports should be written in a manner appropriate to the intended readers. They should be factual, but the degree of detail, structure and terminology should vary according to the requirements and level of understanding of the recipients. For example, management will be interested in broad problem areas, organisational implications, costs and benefits, rather than in the procedural details required by user staff. A report needs to be oriented towards the requirements and interests of its recipients. This does not imply that partial truths or any distortion of the facts are acceptable; on the contrary, because of the sensitivity of people towards computer-based systems, systems analysts have a particular responsibility to be honest in their presentation of facts and recommendations. Trust and respect can be lost very quickly, and are difficult to regain.

Systems analysts should strive to be unbiased and objective in their reporting. While facts, and arguments based on them which support the recommendations, should obviously be included, facts which may run counter to them should not be suppressed. Yet this does not mean that everything must be included or given equal weight. Systems analysts can be selective and use their experience to discuss the facts, but everything that is relevant must be included. Any suppression, seemingly trivial at the time, if subsequently discovered, can cast doubt on the integrity and competence of the systems analysts and consequently on the validity of the system.

Before setting out to make a case for a particular line of action, the systems analyst should first refer back to the terms of reference, checking that all the relevant facts are assembled and that supporting data is accurate. This calls for cross-checking figures for validity with other sources; exploring alternative interpretations of the findings and solutions; and giving support to the recommended solution, not only by presenting facts and arguments for it, but by counter-argument against alternative solutions. This also forestalls subsequent objections and may save time in gaining acceptance of the proposal.

The following is an example of a brief report entitled 'Report Writing': its content describes the principles of report writing within the structure of a report.

REPORT WRITING

B S LEE   July 1978

CONTENTS   *Page No.*

1 SUMMARY

Reports are essential tools of communication for the systems analyst and are often the major contact with senior managers. Reports should be:

- well prepared;
- well written;
- well presented.

The format of reports should be:

- title page;
- contents list;
- summary page;
- introduction;
- findings;
- conclusions and recommendations;
- appendices.

The contents of reports should be:

- at the right level;
- impersonal;
- persuasive;
- precise;
- clear;
- concise;
- comprehensive;
- courteous.

2   **INTRODUCTION**

Terms of reference

To produce a document which would contain advice on preparing and producing reports and which would itself be an example of such a report.

Method of investigation

This report has been written after a certain amount of consultation with other staff and in the light of various publications on writing which are listed in Appendix A. It should be realised that the writing of reports is a very subjective exercise and that the comments in this document reflect the author's personal views rather than some definitive standard.

3.   **REPORT WRITING**

3.1 **What is a report?**

The word 'report' can cover a multitude of documents which vary in size, and in purpose from informing to persuading. In systems analysis a report is usually the formal communication of the reasons for, nature of, results of, and conclusions from, an activity (eg a comparison of hardware devices, a feasibility study, a system review).

3.2 **Preparation**

The report writer cannot just sit down and produce a report; it has to be planned carefully before writing commences and this involves the author in considering the following points:

Questions about purpose

What is the subject of the report?
Why is the report needed?
What effect will it have?
Who will read it?
What work has been done previously?
What period is to be covered?
Where has information come from?
How is it to be checked?
How urgent is the report?

Having considered these questions it is wise at this stage to write a paragraph which summarises the purpose and content of the report and to select a title. This will then guide the selection of material.

Authoriiy for the Report

Usually a report is written by a subordinate to a superior at the latter's request. It is always wise to obtain authority for a report from the people (usually managers) who are able to act on it, and to give copies of the report only to authorised recipients. Unauthorised reports, or reports which fall into unauthorised hands, are often the cause of misunderstanding and distrust and tend to be used as political tools. Another benefit of obtaining authority for a report is that it is incumbent upon the person authorising the report to define what is required. This assists the author to meet objectives.

3.3 **Format of the Report**

The report must be organised so that the readers can easily find their way through it. Basically they will want to answer three questions: what is the problem and how has it been investigated? what has been discovered? what conclusions/recommendations have been reached? The answers to these questions will form the main sections of the report. In addition it is customary to provide a summary page which contains the main points of the report, and to place detailed information, which would hinder the flow of the report, into appendices. The recommended format of a report is as follows:

- title page;

- contents;

- summary;

- introduction;

- findings;

- conclusions and recommendations;

- appendices;

iii

– distribution.

The content of each of these sections will vary from report to report, but some similarities can be pointed out.

Title Page

The title page should be uncluttered and contain only the title, author's name and department, and date. Sometimes a subtitle may be added beneath the title.

Contents

This should include a list of all sections and subsections together with page numbers.

Summary

The summary should not exceed one page in length and should be an attempt to extract and highlight the main points of the report. It should act as a preview for the person who will read all of the report and as a synopsis for the person who wishes to read selectively.

Introduction

The introduction aims to answer the question 'What is the problem and how has it been investigated?' It should therefore contain subsections on the terms of reference, method of investigation, problems of investigation, and anything else which the author wishes the reader to know before commencing the main part of the report. In a lengthy and complex report part of the Introduction may be used to guide the reader through the report.

Findings

This is the main body of the report and will usually consist of several chapters. It should flow easily, and so details like statistics, specifications, etc should be removed to an appendix (with a cross-reference provided).

Conclusions and recommendations

This section is a simple brief statement of conclusions derived from the findings and recommendations based on these conclusions.

iv

Appendices

The detailed information is placed in appendices which should be easily accessed and understood.

Distribution

A list, normally on the last page, of the names of people authorised to receive copies of the report.

3.4 Text Layout

Some organisations have established rules for the preparation and layout of text, and systems analysts must be guided by these. Where such rules do not exist, systems analysts must decide for themselves the layout of the reports and make sure that the typist clearly understands what is required, otherwise inevitably time will be wasted in redrafting and retyping. Some of the points of layout which systems analysts, as authors, must decide upon are set out below:

- the layout should be of a uniform and pleasing appearance with the contents organised so that they can be easily understood;

- a logical arrangement of headings and subheadings contributes to ease of reading, aids understanding and establishes the relative importance of each section;

- a logical paragraph numbering system maintains the relative importance of paragraphs and assists cross-referencing in the text by section and paragraph;

- abbreviations should be preceded by the full name the first time they appear in the text;

- technical words should be defined the first time they are used in the text, or separately in a glossary.

It will also assist typists if the typing requirements are stated before the text, such as:

- stationery size (usually international standard size A4);

- size of both left and right-hand margins;

v

- indenting of headings and subheadings with their following paragraphs as required;

- the use of capitals and underlining for headings and sub-headings;

- spacing, between headings and first paragraph, between paragraphs, within paragraphs and lists;

- space required at the foot of each page;

- rules for typing numbers in words or numerals.

3.5 Writing the Report

Level of understanding

Reports are normally written by a knowledgeable person for the benefit of someone less knowledgeable. The onus is on the writer to be fair and objective, and to present material in an intelligible way.

Write impersonally

It is essential that a report is believed to be objective and one way of achieving this is by the avoidance of personal pronouns and adjectives (ie 'it was decided that' rather than 'we decided'). Once a report appears in black and white, it is considered to be definitive and to have a certain authority. This authority should not be damaged by personal involvement.

Write persuasively

The author must be honest in presenting ideas and should not distort the evidence, but should organise the report to achieve the purpose; objections should not be ignored but forestalled, and the argument should lead to a clear conclusion. The use of techniques like faint praise ('this seems to be an adequate solution, but...') and odious comparison ('this offers a 10% reduction in stock levels, whereas the other system offers a 30% reduction...') is common in persuasive reports.

Write precisely

Vague phrases such as 'a large percentage' ($=51\%$) or 'a thorough survey' ($=3$ out of 10) tend to make the reader suspicious. And suppressed or distorted information will cast doubts on both the report and the author's integrity.

Write clearly

The writer must ensure that the report has a clear structure which can be easily followed, and avoid jargon. Words should be used carefully to avoid misunderstanding and doubt.

Write concisely

Words should only be included if they are essential; facts likewise.

Write comprehensively

The author should set out to answer as many questions as possible that are likely to arise in the reader's mind.

Write courteously

The tone of the report is very important; it should avoid upsetting the reader for the wrong reasons.

3.6 **Style**

The style of a report will reflect the writer's personality and background and it is not possible to legislate about perfect style. Certain observations, however, can be made (culled from writers of 'good' English):

- passages should be broken up into well constructed, logical *paragraphs* which in themselves have a beginning, middle, and end;

- *sentences* should be short (usually no more than 20 words) but not 'clipped';

- a *thread* should run through all sections, and in a long passage/section points should be linked and re-emphasised;

- *words* should be carefully chosen, familiar to the reader, essential and clear; the following points are worth considering:

  - short words are better than long ones;

vii

- adjectives should be used sparingly;
- transitive active verbs are preferable to intransitive passive;
- prepositions and conjunctions should be simple;
- concrete words are preferable to abstract;
- ambiguous words should be avoided;
- unrelated pronouns can mislead;
- emotional words may alienate the reader;
- quantitatively imprecise words should be avoided;
- abbreviations should be explained (at least once);
- words can have different meanings to different people.

To sum up style, keep it brief, and know your reader.

3.7 Use of Diagrams/Visuals

The use of visual aids in reports is strongly recommended for two reasons:

- they break up solid text;
- they can often put a point across more quickly and more succinctly.

One must be careful however not to include too many diagrams, and to put complex tables into the appendices. The kinds of visual that might well be included are:

- tables	classifications;
	reference;
	interpretation;
	frequency distributions.
- charts	bar charts;
	histograms;

viii

Gantt charts;

pictograms.

– graphs   relationships;

ratios;

breakeven points.

3.8 Packaging the report

The first impression created by a report is often a lasting impression, and it is usually formed from the appearance of the document when it is handed to the reader. Packaging of a report is, therefore, very important, and the following rules are worth following.

Cover and Binding

The cover should be smart and have the right impact; the binding should be secure but also allow easy access. If the organisation uses a standard cover and binding, this should be observed.

Typing and Reproduction

Ideally all reports should be typed, on an electric typewriter with a clear typeface, and using the same typeface throughout the report. Printing should be by offset-litho which gives the clearest and neatest finish.

Layout

The layout of the report should be to the organisation's standards but should aim for brevity.

Timing

The timing of the report can be crucial for achieving the right impact, and so should be carefully considered. In a long project it is worth issuing interim reports to retain people's interest.

4. CONCLUSIONS AND RECOMMENDATIONS

Report writing is not easy; it requires a lot of thought and planning, even for the person who enjoys writing.

ix

Reports clearly reflect their authors; they are the author's main contact with senior management; it is worth spending time to do a good job.

Reports should aim to persuade; they must be well presented and well argued.

Would-be report writers are recommended:

- to read extensively;

- to observe the advice given in this report;

- to spend adequate time in producing good reports.

APPENDIX A   FURTHER READING

B Cooper, 'Writing Technical Reports', Pelican
R Flesch, 'The Art of Readable Writing', Harper
T R Gildersleeve, 'Organizing and Documenting Data Processing Information'
E Gowers, 'The Complete Plain Words', HMSO
R Gunning, 'The Techniques of Clear Writing', McGraw Hill.

Distribution

The names of the recipients should be listed to inform the readers of the report of others that are concerned; and also to ensure that each authorised recipient receives a copy.

x

ORAL COMMUNICATION

In most organisations, proposed systems are communicated to management and staff in writing. However, as computer-based systems become more complex and embrace more of a company's affairs, oral presentation is increasingly desirable as a means of ensuring that the details are properly understood by all concerned.

As written communication increases in volume, it becomes more difficult for readers to assimilate, and for the originators to measure the effectiveness of the communication. Reference manuals should not take the place of training courses, though they may be used as material for a training course. A manual is commonly used to clarify specific points when problems arise or memory fails; just as a competent writer may refer to a dictionary to clarify the meaning of a word or to check the spelling. Thus, the use of a manual should be preceded by an oral explanation of its layout, how it should be used, how it relates to the system and to the people who will use it.

Systems analysts must expect to be asked, or, if not asked, to initiate the oral communication of new systems. This is the best way to instil confidence in users. It allows the feedback to systems analysts of listener reaction, so that any unfounded fears, rumours and misunderstandings can be identified and dispelled. It also gives an idea of the rate of assimilation.

Subsequent to the submission of a proposal to management, systems analysts can also expect to be asked to make a formal presentation to a manager or management committee, or to answer questions arising out of the proposal. In either case, the systems analyst may be accompanied by the data processing manager, but it will be the systems analyst who will be expected to make the presentation or answer the questions.

Presentations

Various techniques of presenting information orally can be learned and systems analysts can prepare themselves well for a presentation. Far more important than techniques, however, is attitude. They must:

- be assured and knowledgeable about the subject to instil confidence in the listeners;

- show an interest in the audience, so that each member feels individually involved;

- avoid being too concerned about their own image.

Preparation for an oral presentation needs to be more thorough since memory plays a much greater part. Although all relevant documentation should be available, it is distracting and time-consuming if the systems analyst has to stop frequently to refer to documentation during the presentation.

Systems analysts must also expect to be asked some questions beyond the scope of the documentation: they must then use their knowledge and experience to provide the answers. Where answers are *not* known it is far better to admit this, giving reasons, than to provide incorrect, misleading or inconsequential answers.

The style of presentation depends on the closeness of the relationship between presenter and audience. It is perhaps better to err on the side of formality until a rapport has been established. However, the degree of formality appropriate to written communications (for which the total readership can never be specified with certainty) is quite inappropriate to the normal oral presentation. If it sounds stilted then the audience may judge the thinking to be stilted.

Particular attention needs to be paid, in formulating the ideas to be presented, to the particular interests and the level of understanding of the audience. The stock control clerk, for example, may know that the part number code consists of one alpha and three numeric characters, while the statement that the new system will effect a saving of ten per cent in stock-holding will have little significance: for management, the converse is true.

The structure of the presentation needs to be carefully planned. It is sensible to begin by jotting down a list of all the facts, opinions, illustrations, etc, that are relevant to the presentation; then to select the main points and develop a logical theme; and finally to put the material in a formal structure. There should be an introduction which arouses the interest of the listener and states the purpose of the communication; and, in the case of a proposal to management, the terms of reference. There should also be a conclusion, summarising what has been said, and drawing attention to any subsequent action needed, eg detailed study of particular parts of a written report, authorisation for further development or implementation. The main body should consist of a logical unfolding of the subject matter, with clear links with what has gone before and ('signposts') what is to follow. Frequent summaries of points made are useful and the talk should end on a memorable note. The well-used dictum 'Tell 'em what you're going to say, say it, and tell 'em what you've said' is very applicable to oral presentations.

Having planned the structure for the talk, the next step is to produce a set of notes with key words written down in legible form. If the speaker feels more confident by writing down the details then the subject keywords, the link words, and any particularly telling phrases should be written in larger print, different script, different colour, underlined, or in a particular position on the page. The subject headings are important, but they may be words which, in the written form, would appear in the body of the text, which the eye would miss in the stress of the face-to-face situation. Careful handwriting is likely to help more in giving the variety of stresses used than a neatly typed version.

There may be occasions when a straightforward factual presentation is all that is required, but often a more determined and persuasive approach is needed. In this case, the presentation must really have an impact on the listener. It should begin and end with memorable phrases (which the presenter should have learned in advance); it should be structured so as to gain support from those who are likely to be in favour; it should be convincing (the thoroughness of the study, the logic of the analysis and the likely benefits should be very clear); it should state quite clearly what the presenter wants by way of a decision, why it is wanted, what it will achieve, when it will be achieved, how much it will cost, over how long and what the benefits will be; and above all it should be confidently and enthusiastically expressed.

When actually giving the talk, the presenter is advised to follow these hints:

- make contact with each member of the audience by looking at each one from time to time;

- avoid, as far as possible, distracting mannerisms;

- adopt a natural and comfortable stance and gesture;

- project the voice so that it is natural and audible;

- aim, by choice of language, to express rather than impress;

- lighten the talk by using stories and illustrations that are relevant but not misleading or distracting;

- be very careful to keep to the time allocated, out of courtesy to the audience;

- make use of well rehearsed visual aids;

- allow time for questions and feedback from the audience.

Systems analysts need to realise that they will usually have a far greater knowledge of the subject matter than their audience; that the ideas, concepts, values and procedures with which they are so familiar, will probably be quite new to the listener who will require time to follow and absorb them. Allowance must therefore be made for emphasis of certain points and deliberately slow coverage of some material. Finally, inexperienced speakers are recommended to ask their colleagues to submit themselves to a 'dummy presentation' and then to give their candid opinions of it before the real presentation takes place.

Visual aids

Visual aids have a very important role in any presentation and should be carefully integrated with the spoken material. A good visual aid helps to break the monotony of oral delivery, maintains audience interest, highlights the spoken word, allows time for assimilation, assists retention of information, and can be used to help the speaker to marshall the subject matter; communication experts emphasise that people learn more from looking than from listening ('I hear and I forget; I see and I remember').

There are many kinds of visual aid which can be incorporated into talks:

- *Films* are useful as introductory material but need to be carefully chosen (the presenter is handing over the role to the film for its duration); the requirement for a darkened room and for time to load and unload the projector may be distracting;

- *Slides* are more useful, especially if employed with a daylight screen because they can be made directly relevant to the presentation;

- *Black or white* boards are very flexible but have the disadvantages of not being portable, tending to be dirty to use and requiring the speaker to have his back to the audience;

- *Flannel boards* and *magnetic boards* are useful for building up piecemeal diagrams and are portable but require a lot of preparation for effective use;

- *Flipcharts* are the cheapest type of visual aid and can be used anywhere but they need to be neatly drawn and cannot be re-used if extra material is added during the presentation;

- *Overhead projectors* are the most successful visual aid. They can be used in normal lighting; portable versions can be purchased; the speaker can face the audience; transparencies can be prepared quickly by hand or photocopying, and they can be written over or used for building up a picture.

The most effective visual aid, when describing an object, is the object itself, which the audience can touch and use.

There are certain basic principles which must be followed in using visual aids. They should not be used for their own sake; they must always have a relevant purpose; they should not cause distraction from the main theme; care should be taken not to put too much detail onto visual aids, neither to overcrowd them with too much information, nor to trivialise the subject by using them; it is wise to avoid talking and writing at the same time; and darkened rooms should be used only in absolute necessity. If these principles are followed, visual aids can turn an obscure presentation into an intelligible and interesting learning experience.

Meetings

The systems analyst often communicates ideas in a formal meeting, such as a project committee or a user department working party; sometimes acting as chairman (rarely) or secretary (more frequently) of such a body.

It is worthwhile spending a little time considering the way meetings should be run.

For a meeting to be successful it should have a purpose which is clear to all the members, and it should be of reasonable size, so that they can all contribute. The spirit of the meeting is more important than any set of rules, and participants should be encouraged to see that the aim is agreement rather than conflict: meetings should not be contests in which there are winners and losers. Members should be given time to prepare.

In preparing for a meeting, it is the job of the chairman to decide who should be invited, when and where it should take place, what items should appear on the agenda, and how the meeting is to be conducted. The chairman needs to study firstly the agenda, to ensure that discussion stays on the right lines; and secondly the members, to be aware of possible alignments/conflicts. The secretary must organise the distribution of the agenda, and any papers, and prepare the accommodation for the meeting. The members should read the agenda and papers in advance, and arrive at the meeting with comments and proposals already prepared.

During the meeting the chairman should ensure a prompt start even if some members are missing, introduce the people and the purpose of the meeting, control the timing, progress and degree of participation, and provide feedback or summaries of what has been agreed (this is especially helpful for the secretary). The chairman should act impartially and allow free discussion. The secretary's job in the meeting is to assist members and to take notes of the discussion. The members should aim to contribute constructively. Where there is disagreement it should occur without hostility.

After the meeting the secretary should produce a set of minutes and, having received the chairman's agreement, circulate them to the members. Members who have been asked to carry out some action should do so and the secretary should follow up the action before the next meeting.

The most important aspect of a meeting for the systems analyst is that it is usually an educational situation in which the members are learning about the systems analyst and the possible new system. These occasions must be exploited by the systems analyst to help people to learn, to build up their confidence, and to encourage full participation in discussion.

STANDARD DOCUMENTATION

Much of the systems analyst's communication with other people involves working documents which describe procedures, forms, files, records, input and output documents, programs, etc. These working documents, in the first place, serve as a record and feedback of ideas for the systems analyst's own benefit; subsequently they become the means of communication with others. They are produced as an integral part of analysis and design, and should not be seen as a separate task, to be recorded only after the work has been completed.

There are many reasons why these documents should be standardised in form, layout and size within the organisation. Firstly, standard documentation aids communication between people in the organisation because each person knows what to expect. If a user were to receive different flowcharting symbols and conventions from different analysts, confusion would result. Secondly, standard documents assist in control of the data processing development; several staff can work simultaneously on projects, and completion of documents can be a measure of progress and understanding by the systems team.

Thirdly, standards help with training in that they are themselves an aid to learning because they provide rules for carrying out analysis and design activities; they ensure compatability of work between old staff and new; and they enable people to be productive more quickly. Finally, standard documents aid the systems analyst in analysis and design work; they provide a checklist of facts that have to be gathered and considered; they speed up assimilation of new information; and they promote early transmission of facts from the investigation to the design stage.

The main benefit to communication of using standard working documents is that individual documents are immediately recognisable and the type of information which they contain is already known and expected. For example, a 'computer document specification' will not only be immediately identified as the specification of a document produced or read by the computer, but a systems analyst or a programmer will know at a glance if it is complete. Documentation standards are also a means of communication between past, present and future. Where standard documentation has not been used, the modification of systems designed in the past has caused severe problems of interpretation and understanding. Consequently modifications have often been incorrectly made and taken considerably longer than expected. These problems should never arise with standard documentation.

The standard documents used by an organisation may be created internally or use maybe made of standard documentation produced by external organisations such as consultants, computer manufacturers, or The National Computing Centre.

As the design activity develops, a considerable amount of documentation is amassed. This needs to be organised into a meaningful, coherent system file, for use as:

- a record of progressive fact-recording and design;

- the basis from which various reports and documents will be produced.

A system file consists of all working documents produced as part of the systems analysis and design activities. To be effective, each document must be readily and uniquely identifiable, and filed in a sequence which allows immediate access.

This can be achieved by establishing a method of referencing every working document, which, when filed in numeric or alphabetic sequence, results in a standard pattern of documentation. When standard references are established for the whole installation, a system file becomes readily comprehensible and any of its parts immediately accessible to all technical staff whether or not they have been previously concerned with its preparation. Such a system is illustrated by the table in figure 7.1.

During the development of a proposed system, and when a proposed system supersedes an existing system, the respective documentation needs to be clearly identified. It is useful to set up physically distinct files, clearly labelled with the system title and system reference shown on each document within the file, eg old system/new system/history:

- *old system file*, containing documents relating to the system at the time of the investigation including related analysis documents;

- *new system file*, containing all documents related to the developing or implemented system;

- *history file*, containing all ideas and proposals considered and rejected, a copy of every superseded document, amendment notifications and amendment logs.

These should be designated as permanent master files to ensure their continued integrity. As the new systems file will need to be amended from time to time, it should be centrally maintained for reference purposes and to ensure that all holders of manuals using the documents are informed of any changes. The system file or its contents should not be removed to be used as working documents or included with reports. Instead, individual documents should be photocopied and subsequently destroyed after they have served their purpose, or filed when they become part of a permanent record.

This theme of documentation will be pursued throughout this book.

SUMMARY

As much of systems analysis involves communication of ideas (to and from users, programmers, computer operators, data preparation and control staff, managers, auditors, etc), the systems analyst must develop communication skills. The majority of this communication will be in written form, and the writing and presentation of reports which are both illuminating and persuasive is of utmost importance. The analyst will also be involved in a considerable amount of oral communication when presenting ideas to management or educating users about a possible new system, and when participating in formal meetings to assess progress or evaluate a proposal. Thus oral communication skills are needed. Above all, the use of documentation standards to aid communication must be emphasised. These standards should be an integral part of the systems analyst's 'tools of the trade', assisting both in clear communication and in analysis and design activity.

4 System planning

INTRODUCTION

Before development work can begin in earnest on a new system a number of fundamental questions have to be answered. Some of these will be determined by the computer development policy laid down by a computer development steering committee. Others will be investigated during the initial study (the first stage of the development project) for subsequent decisions by the steering committee. The decisions made will affect the nature of the other stages of the project. It is therefore appropriate to discuss these fundamental questions before examining in detail the activities of systems analysis and design.

A first decision to be made concerns the approach to developing the new system. A number of different approaches have been adopted by various organisations. These range from the design of individual, isolated, small systems to an integrated plan for a total system. The approaches can be problem-oriented or database-oriented; they include varying degrees of user interface with the system. An early decision on the overall philosophy of computer development is clearly essential.

A second aspect of system planning relates to the approach to user involvement in the development of the system. There are clear advantages to be gained from user participation but effort is needed to ensure that such advantages are achieved.

A third area which requires early consideration is whether computerisation can be justified for the system. The aim of the initial study is usually to test the feasibility of the system. However, many organisations either ignore feasibility tests (on the grounds that they are expensive and non-productive and their predictions of cost and benefit are often widely inaccurate) or conduct them at a later stage in the project (possibly on an ongoing basis). At a later stage a more accurate idea of the costs and benefits can be gained because more knowledge of the proposed system is available. Whichever approach is adopted, a decision has to be made to authorise the expensive investigation and design work. This decision will require some formal or informal assessment of feasibility.

This chapter examines various approaches to system development, user involvement, and feasibility assessment.

APPROACHES TO SYSTEM DEVELOPMENT

The various approaches to system development which are worth considering can be categorised as 'individual applications', 'integrated system – bottom up', 'integrated system – top down', 'process (or decision) – oriented' and 'database'.

Individual applications approach

This approach to systems development involves the design of individual computer-based systems to serve specific application areas in an isolated way without regard to possible integration (see fig. 4.1). This approach has

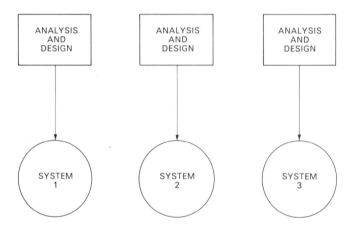

Figure 4.1   Individual Applications Approach

a number of advantages. The systems are easier to analyse, design, program and implement, because they are small and relatively simple. Each system is easy to evaluate in strict cost-benefit terms because it is usually comparable with an existing system and because costs can be more precisely allocated.

Furthermore, this approach provides a way of allowing first-time users to move gradually into computerisation. On the other hand, the approach cuts across many of the systems concepts which were described in the first chapter. It militates against effective information flows in organisations and leads to duplication of both data and effort. It reinforces the departmentation problem and creates difficulties for subsequent linking of systems. The approach is essentially short-term and *ad hoc*, and perhaps only suited in terms of long-term systems development to those organisations which are subject to rapid changes in structure.

Integrated system approach

The objectives of an integrated system approach are:

- to develop an integrated computer-based information system;

- to determine the development sequence for subsystems in a uniform and coherent way;

- to minimise the cost of subsystem integration;

- to avoid the inefficiency of overlapping subsystems.

This approach is illustrated (fig. 4.2). An initial design stage is introduced which produces an overall plan for the design of subsystems in an integrated way.

The integrated system aims to satisfy total information requirements rather than to solve departmental problems, to capture and store data once only (or as efficiently as possible) for all subsequent processing, and to provide up-to-date records of the state of all operations within the organisation. Clearly the design of an integrated system is a complex task requiring careful identification of information requirements, consolidation and rationalisation of data, and standardisation or organisational procedures. It also demands top-management involvement because of its effect on the whole organisation, the considerable cost involved, and the likely need for structural changes to the organisation.

The development of integrated systems tends to be described as 'bottom-up' or 'top-down'. Frequently organisations will use a combination of the two approaches.

Bottom-up approach to an integrated system

Often called the 'evolutionary' approach, this method suggests that the development should initially concentrate on processing of transactions and updating of files at operational level within the organisation. Subsequently the management activities of planning, controlling and decision-models are added. The overall plan therefore reflects the co-ordinated development of

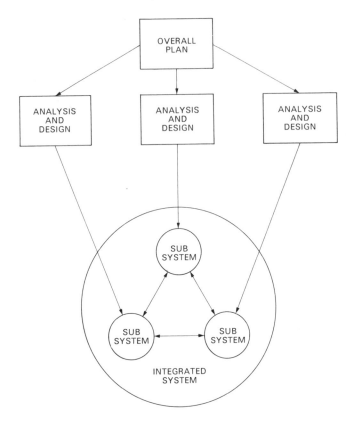

Figure 4.2   Integrated System Approach

subsystems as they are requested. The process might begin with the design in an integrated way of separate applications with their own files to support operational activities; then the files would be physically integrated in a database, thus providing greater facility for information retrieval; next the planning and control activities of management might be developed; and finally the strategic models for corporate planning. This kind of development programme evolves in response to demand and so can usually be more easily justified than the top-down approach. On the other hand, priorities may go astray in the development and the problems of integration can be severe; the bottom-up approach tends to be subject to considerable change as the organisation develops.

Top-down approach to an integrated system

The top-down approach involves the identification at the beginning of system development of a model of the organisation's information needs and the design of subsystems in the light of this model. The integration of the system is planned as far as possible: the plan requires identification of the objectives of the organisation, the functions which serve these objectives, the decisions made by the functions and the actions which result, and the information required to serve the decisions. Clearly this approach demands careful planning and co-ordination and a clear vision at the outset of the ultimate system. It is difficult to justify system development in this approach because its benefits are intangible at the commencement since such a system has never previously been available.

Most organisations work with a compromise between these approaches. They build an overall plan initially (following the top-down approach), then allow the subsystems to evolve in the light of experience gained (following the bottom-up approach).

Process or decision-oriented approach

This approach to system development is in many ways similar to the top-down approach just described but is an approach to individual application design. It seeks to identify the processes with which a manager or supervisor or operative is involved, to analyse the decisions within these processes and to define the data required to facilitate the decisions. The outcome of this approach is a system which collects and stores only the data which is useful to the processes and decisions which have been analysed. Thus this approach can be used within any of the three described above as a means of designing individual applications or an integrated system.

Database approach

The database approach to systems development can be described as bottom-up. It seeks to design the overall system by placing emphasis on the collection, storage, manipulation and retrieval of data which is currently in use in the organisation or can be foreseen as being valuable. The philosophy behind this approach is that it is impossible to anticipate management's requirements for information and so it is more sensible to establish a pool of data which can be readily accessed when the requirements are known. The problems of this are that the wrong data may be stored and that the generalised relationships between data items in the database may be inefficient when compared to the specific needs for processing and retrieval. The database approach involves the systems analyst in identifying and analysing current data in the organisation, and attempting to eliminate redundant data and to include potentially useful data which is not currently used.

The choice of approach to system development is one which must be made at top management level, probably in the computer development steering committee. It is generally a decision which will determine the overall approach of the data processing department to its hardware, software and systems, and will have long-term significance. Changes in technology cause different approaches to be initiated (for example, the advent of cheap mass storage devices made the database approach more feasible) but such changes involve major policy decisions in computer system development. Several attempts have been made over the years to develop computer aids to system analysis and design (eg Autosate, Dataflow, ISDOS, TAG) but they have not been very successful for a variety of reasons. One major reason is that the manual methods which are employed in most data processing departments offer a great deal of flexibility, which none of the automated systems have been able to challenge. It is highly likely that in the not too distant future the computer will be applied successfully to information system development as it has been to so many other areas of human activity.

USER INVOLVEMENT
The extent to which users are involved in the development of computer-based systems tends to depend on the philosophy of the organisation and the individual attitudes of systems analysts. It is generally agreed, however, that the greater their involvement, the more effective the system will be for the users. If a new system is used wrongly, it is because the users do not understand it; if it is used partly (or even not at all), then the system has faults which the users are avoiding.

In either case the system is ineffective. If users feel that a system does not meet their needs, the cause is usually inadequate system investigation, poor perception of the users' requirements by systems analysts, inappropriate design, and lack of user commitment to, and understanding of, the new system. User involvement in the development of systems is therefore of crucial importance: it promotes effective systems analysis and design; it facilitates user understanding and confidence; and it can highlight potential problem areas which are likely to require changes. The case for user involvement is clear; it is less straightforward to achieve.

Conditions for user involvement
Certain conditions must be fulfilled if successful user involvement is to be achieved. First of all, the users must know that their jobs are secure; they are not going to be persuaded to help to design a system which may lead to their own redundancy. Secondly, the users must trust the systems analyst; they must be confident in the analyst's ability and willingness to put their

ideas into practice. Thirdly, the systems analyst must trust the users, being willing to accept the users' ideas (and help them to learn if it is thought that they are not taking advantage of the computer's potential). Fourthly, both users and systems analyst must have a common view of what they are trying to achieve; this probably means that they must be jointly involved in setting the terms of reference for the project. Fifthly, the users and the systems analyst must be in regular contact; they must be willing to listen to each other, they must feel that they have some influence over the design, and they must actively contribute ideas. Finally, and perhaps most important, management must finance the extra time required for user involvement in the various stages of the project.

Methods of involvement

Formal methods of involvement will include user department representatives on project teams and committees. Less formally, there should be regular meetings of working groups to allow staff to familiarise themselves with the development, to provide an opportunity for computer education, and to encourage participation in the design. The users have special knowledge of the existing systems which they can contribute to the design. For example, they will have knowledge, which the systems analyst cannot gain from elsewhere, about:

- exception and error procedures;

- enquiries made of the files;

- task relationships;

- problems of the system.

They will be able to contribute useful ideas on office layout, procedures and techniques, forms design, screen layouts, etc. Above all, they will have ideas about human interactions within the system. The analyst has to remember that the system being developed will eventually have to be worked by the users not by the analyst. Thus the analyst's role ideally is that of a catalyst, enabling the users to redesign their systems; objectivity is offered, and a view of the organisation's total requirements and technical expertise – in the service of the user department.

Systems analyst's role

In order to achieve the kind of involvement that has been discussed above, the systems analyst has various tasks to perform.

Preliminary

First of all, the analyst must assess the possibilities for involvement in terms of company policies, the ethos of the organisation, the commitment of top management, the attitudes of user department managers, the project timescale, and the willingness of the users. If, in the light of this assessment, the analyst concludes that only minimal involvement is possible, perhaps some time should be devoted to educating people about the benefits of participation. Where appropriate, a project plan should allow time for user involvement.

Communication

Once the project is under way, the analyst should examine the communication structure in the organisation and where necessary set up formal structures (eg working parties, briefing meetings, newsletters etc) to ensure that users have information about the system as it develops. In communicating ideas, the analyst should avoid using jargon and make use of standards wherever possible.

Education

At all stages of the project the analyst should concentrate on educating individuals at all levels, using each meeting/communication as an opportunity for influencing attitudes. Departmental managers should be advised to run their own education sessions in the form of briefing meetings at which computerisation, the project, its timescale and implications are discussed. The systems analyst should be available to give computer appreciation courses and should attempt to persuade departmental managers to run such courses. In all of these sessions, the need for user participation in design should be highlighted.

Participation

Both formal and informal structures should be used to allow people to contribute ideas on the old and new systems. All ideas should be welcomed and used wherever possible; where they are not used, an explanation should be offered. Ideally the users should decide on the nature of the new system, with the systems analyst providing information about new methods/techniques and advising on the practicality of user ideas. There should be various decision points in the investigation and design stages at which users give formal approval.

Training

Training sessions should be used to impart new skills and to build up user confidence, not remove hostility (this should have been done much earlier). Supervisors should be encouraged to train their own staff in small groups, with the analyst available as adviser.

Job design

In designing new jobs for user staff, a major aim should be to retain the satisfactory elements of the old job and to remove the unsatisfactory ones. Users clearly need to identify these aspects. They will also have ideas on how the new jobs should be designed – with the systems analyst devoting effort to ensuring adequate job satisfaction.

FEASIBILITY ASSESSMENT

The main objective of a feasibility study is to test the technical, social and economic feasibility of developing a computer system. This is done by investigating the existing system in the area under investigation and generating ideas about a new system. The proposed system(s) must be evaluated from a technical viewpoint first, and if technically feasible their impact on the organisation and staff must be assessed. If compatible social and technical systems can be devised, then they must be tested for economic feasibility.

Assessing technical feasibility

The assessment of technical feasibility must be based on an outline design of system requirements in terms of inputs, outputs, files, programs, procedures and staff. This can then be quantified in terms of volumes of data, trends, frequency of updating, cycles of activity, etc, in order to give an indication of the scale of the technical system. Methods used for investigation, analysis and design of this system are described in the chapters which follow. Having identified an outline system, the investigator must go on to suggest the type of equipment required, methods of developing the system, and methods of running the system once it has been designed.

With regard to the processing facilities, the feasibility study will need to consider the possibility of using a bureau or, if in-house equipment is available, the nature of the hardware to be used for data collection, storage, output and processing. As the technology develops, the range of choice is widening and the complications of batch processing, on line processing, distributed processing, microprocessors, etc, all have to be taken into account.

On the system development side, the feasibility study must consider the various ways of acquiring the system. These include the purchase of a package, the use of a consultancy organisation or software house to design the system and write the programs, a co-operative development with another similar organisation, or in-house development of the system (or a combination of these). Most of these technical possibilities will not be feasible in a given environment and so the eventual proposal will reflect the constraints of the data processing situation in the organisation. It should also reflect the social aspects of the system.

Assessing social feasibility

The assessment of social feasibility will be done alongside technical feasibility. Each of the alternative technical solutions which emerge must be evaluated for its social implications. The needs of various people affected by the proposed system (both directly and indirectly) must be taken into account. Impact on organisation structure, authority, salary levels, group relationships, and jobs should be considered – not just in negative terms (ie bad effects) but positively (ie what can be improved? what contribution can be made?).

The various social costs must also be evaluated; these will include the costs of education and training, communication, consultation, salary changes, job improvements, redundancy payments, and hidden costs like those caused by hostility, ignorance and fear. But primarily the social evaluation should rank the possible technical solutions in terms of the extent to which they improve the jobs and the working environment of those affected.

Assessing economic feasibility

Justification for any capital outlay is that it will increase profit, reduce expenditure or improve the quality of a service or goods which in turn may be expected to provide increased profits. Proposed or developing systems must be justified by cost and benefit criteria to ensure that effort is concentrated on projects which will give the best return at the earliest opportunity.

The determination of development and subsequent operational costs, and of the savings compared with existing systems, can be difficult, especially at the initial study stage, but the quantification of benefits, which management are coming increasingly to expect, is often much more difficult; for example, improved customer relationships arising from the provision of a more informative sales and product analysis for the sales manager, quicker response to customers' enquiries and orders, or faster despatches, invoices and statements.

The technique of cost benefit analysis is often used as a basis for assessing economic feasibility. This type of analysis is not new; it is carried out as a matter of course for many other capital development projects. The factors for evaluation are:

– cost of operation of the existing and proposed system;

- cost of development of the proposed system;
- value of the benefits of the proposed system.

Cost of operation of the existing system

These costs will normally be calculated from cost records. Items to be identified and investigated include:

- manpower;
- materials;
- equipment;
- overhead expenses;
- intangible costs.

Some of these factors are more easily obtainable than others, but they can all be determined within reasonable limits either by interviews with management or by scrutiny of records. Departmental budgets and accounting records are the most fruitful source to reveal the operating costs of the existing system. Costs ascertained by interviews should be confirmed by analysis of the appropriate records.

Manpower costs can be extracted from budgets or payrolls, and it is usual to add an amount relating to the additional cost to the company of each employee such as insurance and pensions.

Materials costs include consumables such as stationery, but stocks and work in progress may need to be considered.

The operating cost of equipment may be expressed as a unit rate if it is established as a cost centre; in this case, it will need to be extended to a period of time, such as annual cost; otherwise depreciation costs, initial or replacement costs of the equipment will need to be estimated.

Overhead expenses are indirect expenses incurred by the company on behalf of all departments, such as rent, rates, power, lighting, and may be extracted from records of cost centres or departmental centres to which they are allocated by the company's accountant.

Intangible costs of the existing system include such things as lost sales, as a result of inappropriate stock levels, or loss of interest because of poor credit control. Such costs must be estimated and included in present system costs. It must be borne in mind also, that the future costs of the existing system may well change. Estimates of growth must be obtained from user departments, and these may affect the viability of the project. For example, the operating costs of the existing system may be increasing rapidly at the time of the evaluation, and if projected to the point in the future at which the proposed system would operate, these costs would be considerably higher.

Costs of operation of the proposed system

The proposed system is likely to include costs in all the areas mentioned above. There are, however, certain other specific cost areas which must be considered in a computer-based system:

- data preparation operators, consumable materials, equipment and maintenance;

- computer running costs, operators, consumable materials, equipment, and maintenance;

- data control staff, equipment and storage;

- system maintenance staff.

Each computer installation will have to have its own method of apportioning these costs. These may all be allocated, including overheads, to the computer as a cost centre and charged out as a rate per hour to company overheads or direct to the user departments concerned. Computer run times for a system, including program maintenance and modification costs, need careful consideration; they have been known to escalate considerably after implementation. Systems analysts must be guided by past experience at their installations, preferably by recorded performance.

Cost of development of the proposed system

The costs of development will be based on the time estimates for the overall project plan. The accuracy and detail of such estimates will depend on the point in development at which justification of the system is required. The estimates at the initial study stage will of necessity be less detailed and accurate than at a later stage of the project. During the later stages, some costs will have become factual, and estimates for the remaining parts of the project should be that much more accurate, being nearer in time and based on later information.

The stages of development of a system to be costed will be those described in chapter 2. Because experience has shown that despite careful planning, deviations occur and unforeseen circumstances arise, some installations include a contingency allowance in the costs to cover this short-coming. This practice is not encouraged by all dp managers because its misuse can negate detailed planning; it should therefore be used carefully and preferably be based on recorded past experience. Particular attention should be devoted to the costs of user education and participation, and project management. Project development cost is a once-only cost, though in some installations it is extended to include any subsequent maintenance and modifications during the operational life of the system.

Benefits of proposed system

The benefits of any proposed system can be considered as falling into two categories:

- tangible benefits;

- intangible benefits (which may be difficult to quantify).

Tangible benefits are those which are readily evaluated in money; they arise out of a direct comparison between the costs of operating the existing and proposed system as net cost reductions or savings. These cost reductions are of the type discussed under 'the costs of operating the existing system' (ie manpower, materials, equipment and overheads). However, direct comparison may show only minimal savings or even increases in costs. A reduction in personnel cost alone is not usually sufficient to justify capital expenditure, unless the reduction is very substantial.

At any one time, a company will have a list of many innovations or projects, against which a proposed system will have to compete. It will be further constrained by the limitations of the company. These include:

- financial limits of cash and credit;

- capacity of management to undertake the heavy load of innovations put forward;

- customers and employees who may be affected by the innovations.

Management are now looking more selectively at the extent to which projects contribute to achieving the corporate objectives of the company within the limitations set out above.

A significant benefit of a computer-based system may be that it can enable the company's financial resources to be used more effectively; for example, in an inventory control system, the computer enables techniques to be used which can provide very large savings in capital, previously locked up in stock holdings, by reducing the quantities previously held, and at the same time offering an improved service.

Another system may lead to improved profit margins by identifying less profitable products; management may then either modernise these production lines or concentrate available resources on the more profitable products to reduce the unit cost.

Intangible benefits are more difficult to estimate and justify, usually requiring the skill of the particular management concerned; in the example used above, given adequate information, production management and the accountant would be able to estimate the effect on profit margins of reducing the unit cost of production. This, in turn, could improve the ability of the company to compete in its markets, but the estimated value of this benefit would require the judgement of the marketing and sales manager. Generally speaking, the computer can be used for lots of activities whose benefits are intangible. For example, greater accuracy, resulting from elimination of boring tasks and more extensive checking, is difficult to evaluate. The computer can make time available to managers for more useful activities instead of record keeping or control, but it is difficult to assess the financial benefit or the likely utilisation of this extra time.

The determination of the value of improved information, or of information in general, is the most difficult task of all. At an academic level it may be said that information is only of value if the possession of information causes the person in possession to carry out some activity which he would not otherwise carry out. The value then is the difference in benefit between the action without the information and the action with it. It is true to say, however, that the manager concerned is in the best position to evaluate this type of benefit.

Presenting the assessment of feasibility

The proposed system must be presented to management with some indication of anticipated performance. The factors which should be included are:

- cost: operating, maintenance, unit, once-off;

- time: response, access, elapsed, resource, cycle, process, management;

- accuracy: frequency, significance and correction of errors;

- reliability: stability, durability;

- security: legal, safety, secrecy, confidentiality;

- flexibility: variability and sensitivity;

- capacity: average, low and peak loads;

- efficiency: performance ratios;

- acceptance: customer, employee, management, shareholders.

Not all of these factors will be important in every system, nor every measurement factor specified with equal precision. For convenience, these may be shown graphically on a measurement scale indicating comparisons between:

- existing;

- acceptable;

- desirable;

- proposed.

On this scale, the systems analyst will be able to record the performance of these factors, as say, average and best. This highlights those factors which need reviewing or justifying in the first place. In establishing what is acceptable and desirable, the systems analyst will try to produce that performance which will gain management acceptance. For this, the systems analyst must know what are the business needs, the views and opinions of individual managers and the capabilities of the data processing facilities available or specified.

A suggested layout for the Initial Study is given in volume 2, chapter 22.

Financial justification

The basic object of any investment is that in return for paying out a given amount of cash today a larger amount will be received back over a period of time; the larger amount should not only repay the original outlay but also provide a minimum annual rate of interest on the outlay. A person investing £100 in a bank would expect to receive back, at some later date, £100 plus compound interest over the period of the investment. If someone borrowed £100 at 10% per annum to invest in shares, the person would aim to receive back, during the period of share ownership, £100 plus interest in excess of 10% per annum. Similarly, to justify a dp project, the investment it represents should provide quantified benefits which are at least comparable to current commercial interest rates.

Because of interest rates, the uncertainties of the future and the effects of inflation, it is generally accepted that money received now is worth more than money received in the future. If £1 were invested today at 5% per annum compound interest it would accumulate to £1.2763 in five years time; therefore, £1.2763 receivable in five years time (excluding the inflation factor) would be worth only £1 today; it must be discounted back to the present at a rate of 5% per annum. While inflation will distort the results, the method still enables valid comparisons to be made between projects of differing costs and cash flows.

Compounding is used to determine the *future value* of present cash flows; discounting is used to determine the *present* value of future cash flows. Just as compound interest tables are available, so also are discounting tables to enable these calculations to be made.

The method employed, of which there are several varieties, is usually based on the use of Discounted Cash Flow (DCF) techniques which are generally recognised as providing a rational means of justifying and comparing investment projects.

DCF methods are based on the consideration of the cash flows of the project and of the timing of these cash flows. In an investment project, money will be spent and received at various times in the future, and the attraction of DCF methods is that the cash flow from alternative investments can be reduced in each case to a single figure for comparison. To do this it is necessary to calculate some rate of discount which will be used to reduce the future sums to their present day value.

The discounted yield method is used to calculate the expected yield, or rate of return from the investment. This rate of return is that which, if used for discounting the cash 'inflows' will make their discounted total equal to the original investment, or, which must be employed to make the net present value (NPV) of the total cash flow (expenditure + receipts) equal to zero.

PROJECT WILL COST £10,000 TO DEVELOP
BUT SAVINGS ARE LIKELY TO ACCRUE

Year 1	=	£310
Year 2	=	£1,000
Year 3	=	£1,000
Year 4	=	£4,000
Year 5	=	£10,000

Year	Net Cash Flow	1st Trial 10% Discount Factor	P.V.	2nd Trial 14% Discount Factor	P.V.	3rd Trial 12% Discount Factor	P.V.
0	− 10,000	1.000	− 10,000	1.000	− 10,000	1.000	− 10,000
1	+ 310	.909	+ 282	.877	+ 272	.893	+ 277
2	+ 1,000	.826	+ 826	.769	+ 769	.797	+ 797
3	+ 1,000	.751	+ 751	.675	+ 675	.712	+ 712
4	+ 4,000	.683	+ 2,732	.592	+ 2,368	.636	+ 2,544
5	+ 10,000	.621	+ 6,210	.519	+ 5,190	.567	+ 5,670
Total	N.P.V.	£	+ 801	£	− 726	£	0
		Rate too low		Rate too high		Rate correct 12%	

Figure 4.3   Discounted Yield Example

This means choosing a rate, discounting the cash flows at this rate, and then examining the NPV to see if it is zero or within acceptable limits. Figure 4.3 shows that in the first trial at 10% the total NPV is too high, therefore the rate is too low; then in the second trial at 14% the resultant rate is too high; and finally a discounted rate of 12% produces a NPV of zero which represents the actual rate of return. The rate of return for this project can then be compared with the company's predetermined minimum rate, and with other projects. To ensure that all projects are justified on a comparable basis there has to be a single centrally determined rate; otherwise varying rates would favour some projects and penalise others. Figure 4.3 gives an example of discounted yield for a project estimated to cost £10,000 to

CASH INFLOWS EXPECTED

	Year 1	Year 2	Year 3	Year 4	Year 5	Year 6	Year 7	Year 8	Year 9	TOTAL
SYSTEM A	6,000	7,000	10,000	10,000	8,000	5,000	2,000	—	—	48,000
SYSTEM B	5,000	10,000	10,000	10,000	10,000	10,000	6,000	4,000	2,000	67,000

DEVELOPMENT COST: SYSTEM A = £25,000   SYSTEM B = £35,000

MINIMUM 20% RETURN REQUIRED

	SYSTEM A			SYSTEM B	
Year	Discount Factor	Cash Flow	Present Value	Cash Flow	Present Value
0	1.000	(25,000)	(25,000)	(35,000)	(35,000)
1	.833	6,000	4,998	5,000	4,165
2	.694	7,000	4,858	10,000	6,940
3	.579	10,000	5,790	10,000	5,790
4	.482	10,000	4,820	10,000	4,820
5	.402	8,000	3,216	10,000	4,020
6	.335	5,000	1,675	10,000	3,350
7	.279	2,000	558	6,000	1,674
8	.233	—	—	4,000	932
9	.194	—	—	2,000	388
		23,000	915	32,000	(2,921)

(Figures in brackets represent cash outflows)

Figure 4.4   Net present Value Example

develop, which is expected to earn benefits in years 1 to 5 of £310, £1,000, £1,000, £4,000, £10,000 respectively. The requirement is to ascertain the expected yield from the investment.

In the *net present value method* the cash value of alternative projects are discounted at the rate selected by the company, and the sum of all the present values gives the NPV of each project for comparison. The example in figure 4.4 in which two systems, A and B, are discounted at 20 % for comparison, shows that system A should be preferred because it has higher resultant NPV than system B.

Various additional mathematical and statistical devices can be applied to the appraisal of investment expenditure for the purpose of allowing for risk, dealing with probabilities, discounting the effect of inflation, and simplifying the calculation of the taxation allowance. Also computer packages for this purpose are obtainable from manufacturers and software houses.

It must be emphasised that the foregoing only lightly touches on the financial techniques used to justify project expenditure, to illustrate the concept of cash flows. While it is unlikely that a systems analyst will be involved in the detailed calculations, there is a need to be aware of the principle involved. Above all, it must be remembered that these techniques depend entirely on the validity of the original estimates, for which systems analysts are responsible.

SUMMARY

Each data processing department has its own approach to planning systems, and the systems analyst has to work within this overall approach. Nonetheless, the implications of the approach taken by the department need to be understood, and this chapter identifies some of the key areas for concern. Firstly, the overall strategy for information system development is crucial as any system under development must fit in with that strategy. Secondly, the extent to which users are involved in the development of new systems will affect their acceptance of change. And thirdly, the assessment of feasibility, both in terms of its timing and its breadth and depth, can fundamentally affect the system development. This chapter argues in favour of an overall strategy, full user involvement, and a broad assessment of feasibility. The systems analyst needs to be fully aware of his department's approach and to have evaluated its consequences in the light of the arguments presented in this chapter.

Part II
System
Investigation

In several of the stages of system development, it is necessary to investigate existing procedures and information flows. The bulk of the investigation will be done early in the development, but it may be necessary as the analysis and design becomes more detailed to carry out further investigation of particular aspects which have been neglected in the earlier stages. This part looks at some of the principles involved in investigating systems.

It begins, in Chapter 5, by examining the objectives of investigating the existing system, and the types of information which the systems analyst is seeking. Chapter 6 explains the various methods which are used to gather this information, with special emphasis on the conduct of interviews which form the main method. One of the most important aspects of an investigation is the record of the results: Chapter 7 describes a number of techniques for documenting existing systems including flowcharting and decision tables (for procedures), data specification, and charts for showing relationships between elements of a system.

5 Objectives of the investigation

INTRODUCTION

Before any work can begin on designing a new system, considerable effort has to be devoted to investigating and establishing what is required. It is necessary to examine current procedures and information flows, to pinpoint problems and difficulties in the existing system, to identify what resources are used, and to discuss with management what improvements are required. Normally the investigation will be governed by the terms of reference for the project, and, once the background has been assimilated, will involve three aspects – management requirements and use of information for decision-making; procedures within the system; and data. This chapter looks at the objectives of the investigation in terms of what information it is intended to produce.

TERMS OF REFERENCE

The terms of reference for the investigation should be provided by the Computer Development Steering Committee or the line manager who has requested the investigation. In some circumstances the systems analyst may draft a set of Terms of Reference and obtain agreement from management before beginning the investigation. There is no point in starting to do work without prior agreement on what the work should be.

The Terms of Reference should include a statement of scope, objectives, constraints and resources.

Scope

This sets the boundaries within which the investigation is to take place. At the same time the analyst's mind must be open to threads connecting the defined area of investigation with other areas in the organisation. If, during the investigation, it becomes apparent that areas are linked by information flow, or in other ways, it may be necessary to seek an extension of the previously defined scope. To be aware of such possible relationships, the analyst is bound to follow the information flow outside the defined scope, having first obtained clearance from the management of the related area.

Objectives

'To establish whether it is possible . . .' may be acceptable as an objective for an initial study but for a full project, objectives should be in quantified form, eg 'to reduce the time from receipt of order to despatch of goods from five working days to two working days, without increasing staff costs'.

Unless the objectives are stated in this way—allowing the achievement to be compared with the objectives—time may be wasted in trying to make improvements in areas which do not critically affect business success.

Constraints

A constraint may be that some particular type of equipment must be employed, a particular existing document will not be changed, or the job of a particular person will not be affected. Awareness of constraints prevents wasting time studying aspects that will not change.

Resources

An indication should be given of the number and type of staff available for the investigation, and of the time within which the investigation is to be completed.

BACKGROUND INFORMATION

The systems analyst should begin the investigation by gathering information which will facilitate the planning of the investigation. This will occupy more time when an investigation is concerned with an area of the organisation with which the analyst is unfamiliar, or when the analyst is newly recruited to the organisation, than when it is an investigation of a familiar function.

The sort of information which is helpful as background includes:

- the history, policies and ethos of the organisation: this sort of knowledge is difficult to gain other than by working for an organisation for a length of time, but sometimes a perusal of annual reports, trade literature, and company procedure manuals will assist the newcomer;

- the economic position of the organisation: the environment of the organisation will affect its internal systems, and so it is important for the systems analyst to have an understanding of the competitive situation, the impact of government policies (especially quotas, grants, contracts etc), and the labour relations climate (trade unions, staff turnover rates, conditions, etc);

- previous and relevant investigations: often work has been done previously in the area with which the analyst is concerned (eg, O & M investigations, job analyses, computerisation projects) and it is useful to read through such documents before starting a new investigation; sometimes books or manuals on particular applications are useful as background material;

- the constitution of the organisation: the systems analyst needs at an early stage to be aware of the functional organisation, numbers of staff in different departments and at different locations, and organisational responsibilities (this is the formal structure, but no less important is the informal structure of personal relationships between particular individuals or groups); other important aspects to discover are the style of management in the departments and the degree of planning.

With regard to a specific investigation, there is need to identify, for the area of the organisation under consideration, who the people are, their tasks and responsibilities, and the reporting structure. This information is obtained from the top downwards. The obvious place to start is the manager who has asked for the investigation. This person may be able to provide an organisation chart, indicating the line of command through which approaches should be made. (It does not follow that the actual responsibility of a given person exactly equates with the responsibility shown on a formal organisation chart. The system may reflect the informal organisation, and any proposed new system must take account of this.)

The organisation chart shows who is responsible for what function and to whom. This indicates, at the next level down, what jobs each person performs. Where possible the attitude of the people from whom information is to be obtained should be ascertained. Individual organisation and individual specialist departments may have their own terminology. It is useful to be aware of this, so as not to appear completely ignorant when these terms are used.

Once the analyst is satisfied that sufficient background for the investigation has been accumulated, the approach can be planned. Usually the starting point will be the departmental managers who will have views on the requirements for the new system, will be able to talk about departmental objectives and information needs for decision-making, and will direct the systems analyst to the appropriate staff for discussion of current procedures.

MANAGEMENT DECISION-MAKING

The systems analyst must aim to discover from departmental managers what they want the new system to achieve. Usually requirements will be expressed in terms of outputs from the computer, their content, frequency of production and usefulness. In order to identify accurately the output information which the manager needs, the systems analyst must help the manager, who may have no previous experience of computers, to talk about his department's objectives, the decisions that are taken, the information required to facilitate the decisions, and the constraints.

Objectives need to be discussed in terms of long-range planning and short-range tactics to achieve the plans. Clearly the long-term objectives are mainly the concern of top management, but they should be reflected in the short-term objectives of the departmental manager. In order to define overall organisational objectives the systems analyst firstly needs to ask questions about expansion plans and methods of expansion in a manufacturing organisation; objectives may well be expressed in terms of increasing market share, increasing the total market, introducing new products, etc. The importance of these objectives is the impact they will have on systems; will there be an increase or decrease in the number of salesmen, products, and branches, in stock levels, in production capacity etc? The second area of questioning is concerned with the methods and structures which have been established to ensure the achievement of long-term plans; the systems analyst needs to find out who is responsible for achieving long-term aims with regard to such things as sales volume, sales promotion, product development, product quality, product costs, production control, purchasing, manpower planning, etc.

Having established in outline the position of a particular department in relation to overall organisational objectives the systems analyst can then discuss the specific objectives of that department with its manager. Points which help to define the objectives of a department would include responsibilities (up and down the hierarchy), task requirements, performance measurement, delegation, likely changes, and problems of current systems. The systems analyst needs to establish with the manager specific areas of responsibility, (including decisions which have to be made) and the information required to undertake the responsibility and make the decisions. For example, a sales manager might identify responsibilities concerned with selling, pricing, discounts, order processing, market research, advertising, catalogues, product introduction and withdrawal, stock allocation, and forecasting. Each of these could be expressed as an objective, with an indication of how performance against the objective is measured, and what information is needed to facilitate both good performance and fair measurement of performance.

So far the examination of objectives has been considered at the level of what should be rather than what is. It is important that the existing mode of operation is also considered, and a discussion of the problems highlighted and improvements requested by the departmental manager will help the analyst's understanding of departmental objectives and information flows.

PROCEDURES

In investigating current procedures in a user department, the systems analyst should aim to find out in a comprehensive way what each member of the department does, at what times, in what way and for what purpose. This involves consideration of normal procedures for dealing with particular operational documents, control procedures for ensuring the accuracy and security of such processing, and exception procedures for dealing with errors or unusual incidents or abnormal loads. In each of these areas the analyst must discuss the problems faced by the clerical staff in carrying out the required procedures, and must try to identify the resources which are allocated to each part of the procedures. This includes not only staff time but also equipment, accommodation and supplies.

Knowledge of existing resources indicates the extent to which they may be used to fulfil objectives. This can help to minimise cost and disruption in the department. There may also be skills and equipment, which could be put to better use in some other function. The analyst should report on such possibilities.

The bulk of clerical procedures will be concerned with handling data (as described in the next section) ie producing outputs from inputs and keeping master files up-to-date. It is necessary also to identify activities such as answering the telephone and writing letters in order that a full picture can be gathered of how staff time is divided between the various activities. The pattern of activity in a particular procedure is important and information on the frequency, peaking and growth of procedures must be gathered.

The analyst will find that the major part of the investigation is devoted to two particular aspects of clerical procedures: exceptions and problems. The exception situations for which the analyst needs to watch, include:

- peaking situations: there will be occasions in the system cycle when abnormal workloads will be experienced, these can be due to seasonal variations, holiday periods, year-end accounting activities, stock-taking, etc;

- reporting: the production of reports for management can be irregular and time-consuming (especially if *ad-hoc* reports are frequently requested);

- errors: even in the best system, errors will occur, such as incorrect code numbers, controls which do not balance, negative stock values, etc;

- missing data: an important exception situation arises when input documents contain omissions, or when there is a mismatch of an input document with a master file;

- special situations: all special procedures, such as action on priority transactions or procedures for new customers or products, should be specified.

The kinds of problems in clerical procedures that the analyst has to investigate are many and varied. Normally the analyst will discover or be told about delays in receiving documents, excessive document or staff movement, arrears in processing, too much copying and too many records. These are not so much problems as symptoms for which the analyst has to identify the cause; and usually there are several contributing factors which need to be examined. If the analyst is able to identify the cause of problems then significant and rapid improvements may be possible with current procedures. However, the main aim is to learn about current problems so that they do not reappear in the new system. Most problems in clerical procedures can be grouped into the following categories:

- workload, which can be unevenly distributed, poorly organised, too dependant on overtime, or excessive because of understaffing;

- documents, which can be poorly designed, non-standard, require too much in writing, and too many copies;

- controls, which can require too much (or too little) accuracy and precision;

- office layout, which can hinder work flow, prevent adequate supervision or cause consecutive instead of concurrent actions;

- training, which is often inadequate, or may lead to overspecialisation;

- equipment, which can be out of date, inefficient or frequently inoperative;

- environment, which can cause slow working, interruptions or distractions.

Where the existing system is already computer-based, the investigation will need to extend into the computer area. Fact-finding will be considerably eased where the existing system is well-documented, particularly if it adheres to an installation standard.

DATA

Data used in clerical procedures can be placed in two categories, input/output and files.

Input may be in the form of a document, or may be received by word of mouth. It is important that all inputs, from whatever source and by whatever means, are fully recorded. This means either asking someone to complete blank forms or photocopying existing ones. Since more information is required than simply the content of the document, it is advisable to use some form of checklist. The Clerical Document Specification described in chapter 7 provides a useful discipline. It will usually be found convenient to keep this with a sample of the document facing it.

The same considerations apply to output, though additional information is required, eg names of recipients, their location, the purposes for which the output is used, and how often.

There may also be some intermediate documents within the department which serve only as a means of preparation of, or proforma for, an output. The Clerical Document Specification provides an equally useful checklist for these.

There are essentially two types of non-computer file: master records or accumulations of various documents. (It is, of course, increasingly likely that any existing system will include some computer files.) Master records frequently take the form of cards, but may be either loose-leaf or bound books. It is particularly important to know the use to which these are put. Document files may be divided into two categories. The first contains a series of documents of the same type, eg copy invoices, copy orders; the second may be simply a collection of various forms and correspondence.

Quantitative information about documents and files is extremely important because of its impact on the physical design of the system. The volume of particular documents, the frequency of their use, their growth rate, etc, will determine file sizes; the size of data items and their interrelationships will affect file structures; and an analysis of file utilisation will assist in deciding on file organisation and access methods.

SUMMARY

In the investigation stage of a system development project, the systems analyst needs to be absolutely clear about what is being investigated; this should be covered by the Terms of Reference. It is also necessary to be familiar with the system environment and terminology; this involves research into the background of the system. Finally, an investigation should be conducted into three aspects of the system: first, the management decision-making requirements; second, the procedures involved in the current system (especially error and exception procedures); and third, the data used as input to or output from the system and the data stored in files.

6 Methods of investigation

INTRODUCTION

There are several methods of gathering the sort of information that has been described in the previous chapter, and normally the systems analyst will use all of them in an investigation. They include observation, record-searching, special-purpose records, sampling, questionnaires and interviewing. This chapter describes them in some detail.

The analyst's approach to the investigation will influence the use of the various methods; it is important that the approach is appropriate to the situation under consideration. The approach should initially be formal, with approval sought from staff at higher levels in the organisational hierarchy before contact is made with staff reporting to them. Confidentiality and honesty are important in all dealings and at all levels, so that the interviewee feels that the systems analyst is reliable and objective. The analyst should encourage participation in the investigation by welcoming all information which is offered and especially by recognising the superior knowledge of the user of the existing system.

As far as possible, the analyst should keep appropriate people informed about the progress of the investigation, both out of politeness and as a means of retaining interest and involvement in the task. The investigator should always be grateful for the help received from users.

There now follows a description of methods of investigation. They should not be looked upon as *independent* activities but as *integrated*. Observation will take place during interviews: questionnaires may be completed as part of an interview; records may be inspected during an interview or any other discussions. Each method complements the others.

OBSERVATION

Many important clues can be picked up by the trained observer. For instance:

- office conditions: noise, lighting, temperature, ventilation;

- layout: whether the staff have to do a lot of walking about, whether access to desks, filing cabinets and other equipment is easy or difficult;

- office furniture: is there enough filing equipment or are files stacked on the floor? Is there a standard for filing equipment or a random assortment?

- supervision: do the staff seem to be getting on with their work or are they chatting or reading magazines? Are there frequent interruptions?

- work load: do the piles of documents waiting for attention stay constant or do they fluctuate from one part of the day or week to another?

- bottlenecks: are there operations at which there is constantly a heavy load of work waiting for attention?

- pace of work: this requires training, and the trained observer can recognise a slow, a normal, or a fast pace of work;

- methods of work: this is not something gained from casual observation but needs deliberate attention. It is unlikely that this knowledge can be gained without the conscious co-operation of the person being observed, and the analyst should always arrange to be introduced to the person and give an explanation of what is going to be done and why. Some interviewing may also be necessary to explain the logic behind some of the operations.

The clues which are picked up should always be followed up by questioning in interviews.

RECORD SEARCHING

The main purpose of a record search is to establish quantitative information – volumes, frequencies, trends, ratios. It will also help to establish how much reliance can be put on the estimates given by the staff or the management of a department. It may also indicate whether the departmental objectives (eg the time from receipt of order to despatch of goods) are being achieved and whether information needed for decision-making is available when required. It may disclose exceptions and additions to the information obtained from interviews – such as documents not mentioned, or received and filed without any action.

Among existing records may be found written procedures (perhaps from an earlier O & M investigation) or job descriptions which can of course be extremely reliable for background information.

SPECIAL PURPOSE RECORDS

Sometimes the existing records do not supply the information required, and the only way of obtaining reliable information may be to install, for a limited period, special-purpose records. The missing information is likely to be quantitative, concerned with volumes, frequencies, trends or ratios, or it may relate to management information requirements. The kind of information that one might wish to gather in this way would include, for example, the time at which documents arrive in a department, the number of times files have to be consulted, the volume and frequency of telephone conversations, the number of queries to management, the type of enquiries from management, etc.

Account needs to be taken of the fact that keeping the special records will be additional to the normal work done, and therefore must affect, however slightly, the volume of work being processed. To minimise this effect and cause as little ill-will as possible, the record should be simple: it may consist of doing nothing more than making a mark on a line or in a column whenever a particular activity takes place, or making an extra carbon copy of each document produced. To avoid the suspicion that it is a permanent addition to the work-load, a specific duration should be set, ie whatever minimum time will give a representative sample of what is required to be known.

SAMPLING

Where there is a high volume of documents or a number of recurring activities, it may not be necessary or practical to measure the total number. The economical way may be to take a sample. The safest way is to use random numbers, but it can be acceptable to count every 'n'th document or take a reading every 'n' minutes, so long as 'n' does not coincide with some cycle in the occurrence of the types of document or activity.

The sampling technique can be used to reveal volume, frequencies, trends or ratios. It is also commonly used as a method of work measurement.

QUESTIONNAIRES

This form of fact-finding must be used with great caution. Designing questionnaires is a specialisation within the general subject of form design. Considerable skill and field testing are required to avoid confusion, ill-will, and misinformation. It is only worth spending the time and effort required if the information is needed from either a large number of respondents or a number of remote locations.

The important considerations to bear in mind when designing a questionnaire are its purpose (what specific information it is intended to gather), its recipients (their level of understanding, intelligence and interest) and its timing. The questions must be carefully phrased so as not to be ambiguous, not to lead the recipient to false answers, and not to prevent relevant answers from being included. A covering letter, making clear the purpose of the questionnaire, addressed to a specific person, and enclosing an envelope for return are essential to achieve a good response. The questionnaire itself should normally be in three sections:

  i   Heading section, which will provide brief information about the
      questionnaire and will allow the respondent to fill in the date, name
      and title;

  ii Classification section, which will include the data to be used for
      analysing the replies (eg, sex, age, location etc);

  iii Subject section, which will be the specific questions forming the main
      body of the questionnaire.

INTERVIEWING

Interviews are by far the most common and most satisfactory way of obtaining information, particularly to obtain information about objectives, constraints, allocation of duties, and problems and failures in the existing system. To be effective and economical, interviews need to be well planned.

Whom to interview: this depends on who has the required information. It may be advisable, even though time-consuming, to go down through a hierarchy to arrive at that person.

When to interview: there may be a most desirable sequence, based on the kinds of information to be collected, so that one interview can build on information already collected. Since the analyst is generally in the position of asking a favour, no attempt should be made to dictate when the interview shall take place; it needs to be arranged so that it causes the least disturbance to the work and to the personal arrangements of the person being interviewed. At the same time, the systems analyst must not be put off too easily; project requirements have to be met.

What to ask: to make the best use of the time available, it is important to obtain prior knowledge of the duties and responsibilities from which a check list of questions can be drawn up. It is important to ask questions at the right level; for instance, one would not ask the general manager how many digits there are in the order code or the junior clerk why Jones gets 10 % discount and Brown only 5 %.

Where to hold the interview: often the only available place is the normal work-place of the person being interviewed. This can be an advantage, since the interviewee will feel more at home and additional information can be obtained from observation. Interruptions may tell a lot more about the problems than the interview itself. Exceptionally, if the subject to be discussed is not directly connected with the job being done, or if the interviewee normally works in a noisy environment or mainly on the telephone, then it may be desirable to use a separate interview room.

In planning the interview, the analyst must always be considerate to the interviewee, and decide what is most convenient/easiest for the interviewee. The art of interviewing is being able as early as possible to make the interviewee feel at ease; and this is done by helping the interviewee to feel that ideas elicited are important, by establishing a relationship of confidence, and by commencing the interview with topics that are easy to talk about.

Beginning the interview: the entrance and the first few words, especially at a first interview can be crucial. A late arrival, or an excess of formality or informality can cause quite unnecessary problems. The degree of formality is a matter of experience and judgement. Punctuality is essential.

The interview: it is important to remember that an interview is an interruption, perhaps a welcome one, but this will not be known until contact has been established. The other person may not remember the arrangement that has been made and may not want to remember. The systems analyst should therefore begin by an introduction and explanation of the purpose of the visit.

It is important to know what information it is hoped to get, and to aim not to leave without it. This does not mean reading questions out from a list; successful interviewing usually consists of letting the other person talk, with occasional prompts in the required direction. Unless the analyst has a very unusual memory, it will be necessary to take notes, and as a matter of politeness, ask if this is acceptable. All relevant comments need to be recorded with opinion carefully distinguished from fact. Ideas for improvements should be encouraged and noted, but without criticism or commitment.

Listening is an art required by the systems analyst. Care should be taken not to cut the interviewee short. Questions should be framed in an open way so as not to invite the answer 'yes' or 'no', but rather to encourage the interviewee to talk. The analyst should also take care not to use leading questions ('Is this because . . .?') and not to ask multiple questions at one point, only one of which will be answered; most questions should be of the 'how? when? why?' type and should be encouraging. It is very useful for the analyst to have a grasp of the technical vocabulary appropriate to the interviewee and to use it. But, above all, whilst computer jargon must be avoided, it is a good time to introduce users to the correct use of terminology used by data processing staff. As the interview progresses, the analyst has got to keep one step ahead of the interviewee and be able to steer the interview in the appropriate direction. At the same time informative diversions should be noted. A good interview is one in which both parties feel relaxed and achieve a sense of rapport. This will not be helped by bored looks or argument or biased questioning from the analyst. The analyst must be sensitive both to the needs of the interviewee and to the assumptions and attitudes expressed. Often more can be learned by reading between the lines of opinions expressed than from the opinions themselves. It is said that 'listening is a process of self-denial'; the systems analyst must learn this process.

Clearly a lot of information is learnt in an interview situation from non-verbal aspects of communication. Facial expressions, eye movements, gestures and physical proximity provide signals of the interviewee's attitude to the interview or to points which arise. The interpretation of these signals is subjective and sensitivity to them is innate, growing with experience, but the analyst must realise their importance. Even before the interview commences, the interviewee will have formed a perception of the systems analyst (via letters, memos or reputation). This perception will be modified on meeting (influenced by appearance and initial conversation), but will tend to consist of a simple, possibly inaccurate, categorisation of the analyst. This categorisation will determine the social technique which the interviewee adopts.

The analyst has to create (via good preparation, good organisation, and appearance) an affinity with the interviewee which quickly breaks down any barriers and allows a relationship to grow. Thus the systems analyst must carefully select a social technique to suit the interviewee and adjust it so as to achieve rapport. This rapport is most easily achieved when there is a clear channel of communication, some degree of acceptance, and a smooth pattern of interaction.

The onus is on the systems analyst to show a genuine interest in the interviewee, by a warm and friendly manner and by expressing interest in the interviewee's problems and terminology. The systems analyst must keep the other person involved in the interview by agreeing and admiring, encouraging participation, and ignoring displeasure or lack of response. This demands that the systems analyst can deploy a wide range of social techniques which may be used flexibly, energetically and smoothly.

The systems analyst must always be honest in interviews, avoiding false assurances and issues beyond control. In approaching the user, the analyst should make allowance for misconceptions and misinterpretations; and be analytical in determining what really happens. Above all, the analyst should not rely on user management support unless it has been explicitly given.

Concluding the interview: before leaving, the systems analyst must check the important facts that have been noted. If the analyst's writing is legible, then the other person can be invited to read what has been written; otherwise, a record prepared afterwards should be passed over for verification. The systems analyst should also obtain a copy of any user documents which have been mentioned. The copies should have sample entries completed. It is important to leave on a pleasant note, leaving the way open for a possible return. There will often be facts to obtain which have been neglected, or a subsequent visit may be used to start sounding out views on ideas for improvement.

After the interview: all notes should be read through and expanded so that they are intelligible. If they are in random order, they should be revised into a more useful order before further work is commenced. It may be necessary to rethink the outlines of other interviews which have been planned previously, either because information has been gathered already which was the subject of a subsequent interview, or because points have arisen which change the emphasis or require confirmation and cross-checking elsewhere. The latter is very important, the account of one person about what happens in another section should not be accepted without an on-the-spot check.

SUMMARY

There are several methods of gathering the information required about a current system, but each method has particular advantages and disadvantages. Observation of the work situation will provide clues to problems (eg bottlenecks) and atmosphere, but such clues have to be followed up in detail. Record searching, use of special-purpose records and sampling will give quantitative information about the system which facilitates sizing of any new system, and may also point to areas of difficulty which are being experienced. Questionnaires allow the gathering of a restricted range of information from a large number of people; they can be used to ascertain people's attitudes or to collect quantifiable data about a system (eg number of staff, type of machines used, etc). All of these fact-finding methods need to be supplemented by more detailed discussion of points in an interview situation. Identification of user requirements, decision areas, objectives and responsibilities for certain procedures, can only be achieved by interview. Thus interviews are the most important fact-finding method and need to be well-planned and well-conducted. Once the facts have been gathered they need to be adequately documented.

7 Recording the investigation

INTRODUCTION

Although the investigation, fact-recording, and analysis are treated in this book as separate activities, this does not mean that the process of recording and analysing the facts is to be deferred until the investigation is complete. As each set of facts is ascertained it needs to be conscientiously documented and a preliminary analysis carried out to reveal any areas that need further investigation.

The NCC Data Processing Documentation Standards provide a tried and tested means of recording the investigation. This chapter examines the forms involved and how they can be used to aid the systems analyst in documentation. It looks first at a method of referencing and cross-referencing documents, and then at various methods of recording un-structured facts and agreements, procedures, data, and relationships.

DOCUMENT REFERENCING

The NCC's document referencing scheme provides nine main areas into which a system file can be divided. They are defined in sufficiently broad terms to accommodate any document arising from the activities of analysis and design. Whilst it is not necessary to have documents in all of these areas, the headings do indicate categories of relevant information. The nine main headings are:

- 1 background;
- 2 communications;
- 3 processes;
- 4 data;
- 5 support;

111

- 6 tests;
- 7 costs;
- 8 performance;
- 9 documentation control.

The numbers are used on the standard forms themselves to indicate which area the form belongs to, and can be further subdivided if required. Details of NCC's advised subdivisions are given (fig. 7.1). This number is placed in the box headed 'Document' which can be found at the top of all NCC systems analysis forms. Two further boxes are headed 'System' and 'Name': these usually contain abbreviations or mnemonics allocated within the computer department to link items. There are also boxes for the document title and sheet number, to be filled in as required.

A description is provided to indicate which types of information should be filed in each of the listed areas:

Background (Document Reference 1)

The most important information at the outset of the project is the Terms of Reference. There may also be reports of previous studies of this system, or of similar systems in other organisations. Existing documents may be regarded as part of the background to the study. One typical piece of background information is the organisation chart. This may need to be at two levels, one for the complete organisation and another for individual departments being investigated.

Communications (Document Reference 2)

All the documents referring to decisions made at meetings during the design, development and generation of the system should be filed here. For example, records of meetings made on Discussion Record forms, correspondence and documentation such as the User Manual can be recorded in this area.

Processes (Document Reference 3)

This part of the system file contains details of procedures and processes to be used to meet the objectives set out in the Terms of Reference. These can include a broad systems outline, details of clerical procedures, all the stages of systems design, and a detailed specification of the system. Each level (except the last) can be broken down into a further level. Various techniques are used to help the systems analyst in the recording of procedures and processes. The most common aids are flowcharts and decision tables.

Data (Document Reference 4)

Data consists of the contents of inputs, outputs and stored files, held on documents or on magnetic media. Useful forms for recording the data

SUMMARY OF SUGGESTED FILING REFERENCES

Filing Ref.	Name/Definition
1	1 BACKGROUND. Terms of reference, objectives, constraints.
	2 COMMUNICATIONS. Information gathered concerning purpose and scope.
2.1	Discussions, meetings.
2.2	Correspondence.
2.3	Associated documents, eg user manual.
	3 PROCESSES. Information about methods and actions.
3.1	Overview.
3.2	User-clerical procedures.
3.3	Operations – preparatory processing, distribution of results.
3.4	Computer process-organisation.
3.5	Computer process-details.
	4 DATA. Information about data used or produced.
4.1	Clerical data.
4.2	Source data files.
4.3	Output data files.
4.4	Stored data files/logical data base.
4.5	Source data records.
4.6	Output data records.
4.7	Stored data records/item groupings.
4.8	Internal data.
	5 SUPPORT. Information supporting processes and/or data.
5.1	Analyses and interactions.
5.2	Data item usage.
5.3	Hardware, software, other supporting facilities.
	6 TESTS. Information about action taken to prove the design.
6.1	Specification of test requirements (including data).
6.2	Test plans.
6.3	Test operations (as distinct from system operations).
6.4	Test logs.
7	7 COSTS. Information about equipment and development costs.
8	8 PERFORMANCE. Information about timings, volumes, growth, etc.
	9 DOCUMENTATION CONTROL. Information about use of documentation.
9.1	Copy control.
9.2	Amendment list.
9.3	Outstanding amendments.

Figure 7.1   Filing References

content of any document or file, together with its essential physical characteristics, include the Clerical Documentation Specification (fig. 7.8).

Support (Document Reference 5)

Details of interactions, data item usage and available hardware, software and supporting facilities are recorded in this area, together with reference to or details of other analyses made of the existing system. This area is also to be used for any information which is ancillary to the processes being studied which does not fit into the references 3 and 4 area definitions.

Tests (Document Reference 6)

Thought must be given to how the system can be tested. Information must be gathered about the actions to be taken to prove the developed design; the results of any tests carried out must be filed away for future reference. This can best be done by the use of a separate tests file.

Costs (Document Reference 7)

The costs area should contain documents recording the cost associated with both the existing systems and the proposed new system. Costs should take into account both manpower and equipment, and should follow established accounting procedures within the company. Care should be taken to ensure that the same accounting procedures apply to both clerical and computer departments. If not, details of the difference, together with an explanation of the reasons behind them, should be recorded here.

Performance (Document Reference 8)

This area should contain those documents relating to the timings, volumes and growth of the system and its constituent parts. Details proposed earlier, either in the Terms of Reference or systems specification or on other documents, should be cross-referenced here, as should any relevant test data. This part of the file may be used long after systems development is finished to record performance and to compare existing with past performance.

Documentation Control (Document Reference 9)

Finally an area is required where documents recording the current state of the documentation are kept for ease of reference. Information stored here should include the whereabouts of copies of documents whose masters are in the file, the amendments which have been incorporated together with their history and also details of any outstanding amendments which are to be made as soon as resources are available. Any notes on the use of the filing system can also be kept here.

Experience has shown that some or all of these nine areas will suffice for filing the documents associated with all but the very largest systems. It may be that in a particular system some other arrangement will prove more practical, but consistency between the documentation files of different

Discussion Record	Title		System	Document	Name	Sheet
			MS 10	*2.1*	*F 150*	*1*

NCC

Participants *MRP - Systems* *JBB - Programming*	Date *11/7/77*
	Location

Objective/Agenda *Validation of PR-FCODE*	Duration

Results: Cross-reference

Factory code (PR-FCODE) has value range 1 to 4.
Highly unlikely that this will be extended.
Validity tests may therefore use literal
expressions in preference to table look-up.

S21
Author

Figure 7.2 Discussion Record

systems should be considered before introducing another arrangement. Will the new arrangement suffice for all existing and future systems within the company? Will it contain specific areas to record all the details listed above? These questions must be asked before developing any referencing and filing system.

UNSTRUCTURED FACTS AND AGREEMENTS

During the investigation a lot of information is gathered as a result of interviews and discussions. Notes should be taken during an interview, but usually it is necessary to distill the essential facts afterwards. Some of the facts are best recorded in narrative form; others lend themselves to other methods, such as flowcharting and document specification. The notes should be turned into an appropriate record for future reference by the systems analyst or other members of the project team. The record can also be used for verification of facts by the person who provided the information. Both of these communication processes will be made easier if the record is in a standard format.

For the unstructured facts and agreements which need to be expressed in narrative form a Discussion Record form is recommended (fig. 7.2). This includes space for mention of the people involved, the objective or agenda of the meeting, its date, location and duration, and cross-referencing to flow-charts or document specifications which have been produced along-side the discussion record.

PROCEDURES

It is usually easier and more intelligible to record identified procedures in diagrammatic rather than in narrative form. The two main methods of doing this are flowcharts and decision tables. Narrative is occasionally necessary (for example to expand on the flowchart or decision table), but it should always be directly linked (ideally on the same sheet) to the flowchart or decision table.

FLOWCHARTS

A flowchart, using words to indicate a sequence of events, differs from normal narrative in that it does not use sentences. The words are enclosed in symbols (linked by flow lines) with conceptual meanings indicated by their shape. Flowcharting is the most common method of describing procedures in a computer-based system, whether for the benefit of a line manager, a clerk or a programmer. The advantages of flowcharts over narrative are that:

– they show logical interrelationships clearly;

– they are easy to follow;

- they allow tracing of actions which depend on conditions;

- they can be produced in a standard way and so allow several people to work on them simultaneously;

- they are useful to the systems analyst for experimenting with different approaches to a particular problem.

These advantages begin to decline as the flowchart increases in size and complexity and it becomes difficult to amend the logic. The two major disadvantages of flowcharts are: the difficulty of tracing back from actions to conditions; and the requirement to maintain a consistent level of detail in order to avoid confusion.

Flowcharting symbols

There are several different sets of standard symbols for flowcharting, the main ones being those developed by ASME (American Society of Mechanical Engineers) and ECMA (European Computer Manufacturers Association). ASME symbols are used mainly by O & M and work study officers, ECMA by data processing staff. Neither of these is suitable for both clerical and computer procedures (ASME, for example, includes no decision symbol, and ECMA no movement symbol).

A set of symbols (fig. 7.3) has been developed by NCC to meet the needs of both clerical and computer procedure flowcharting. The set of NCC symbols is small in number, each symbol representing a concept rather than a device. (If devices were used more than 30 symbols would be required and this would always be changing as new devices are developed.)

To indicate the device being used for input/output or file, a bottom stripe is used and the name of the device is written in; for example:

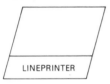 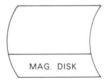

Cross-referencing of processes (to other flowcharts or decision tables) and files, inputs or outputs (to their specifications) is used by writing the cross-reference in a top-stripe of the symbol; for example:

Symbol / Type of Chart	System Flowchart Interactive System Flowchart and Clerical Procedure Flowchart	Computer Run Chart	Computer Procedure Flowchart
1 (rectangle)	All operations or procedures		
2 (diamond)	All decisions		
3 (storage symbol)	Storage media, permanent or temporary	Computer Backing Storage	Not used
4 (parallelogram)	Documents, cards, paper tape, displays, etc.	Data passing between computer and non-computer parts of the system	Not used
5 (circle)	Connector, showing continuity between symbols where it is not possible to join them by a flowline		
6 (terminator)	Terminator, showing entry to or exit from a procedure		
7 (triangles)	Data moving from one location to another	Not used	

Figure 7.3   NCC Symbols

Principles of flowcharting

It is essential to keep the flowchart clear and easy-to-follow by:

- marking the start and end points;

- using standard symbols;

- avoiding crossed flowlines;

- using simple decisions (yes or no);

- working in a consistent direction.

The logic should be thoroughly checked to ensure that no actions are missed or wrongly repeated. The flowchart should show a consistent level of detail (by bearing in mind the use of the flowchart). Finally it is important to verify the completed flowchart by passing data through it and checking that the result is as expected (known as dry-running).

Levels of flowcharting

Flowcharts can be produced at different levels of detail. A System Flowchart which outlines a complete system is shown (fig. 7.4). The symbols are arranged in columns to indicate the function or department within which the activity or entry represented by the symbol occurs.

It is a general convention of flowcharting that flowlines go from left to right and from top to bottom; any exceptions to this convention are indicated by arrowheads on the line, such as the one entering immediately before symbol 5 (fig. 7.4). This is to be distinguished from the closed arrowhead, represented by symbols 13 and 15, which indicates physical movement of the medium containing the information. Note that this physical movement symbol could be used between symbols 5 and 6, but it would be redundant, since the movement is clearly signified by the flowline moving from one column to another.

Where a document containing a lower level of detail is available, relating to one of the symbols on this systems flowchart, cross-reference is provided to that document in the top stripe of the symbol. The Clerical Procedure Flowchart (fig. 7.5) is a lower-level representation of symbol 2 in the System Flowchart (fig. 7.4).This particular flowchart uses the columns to differentiate between the different documents and files used in the procedure. The same procedure could be drawn in a different way to emphasise department activities, or as a single main flowline to emphasise processes. The choice as to how the columns are to be used is one to be made by the analyst, depending on which format brings out most clearly the essential factors.

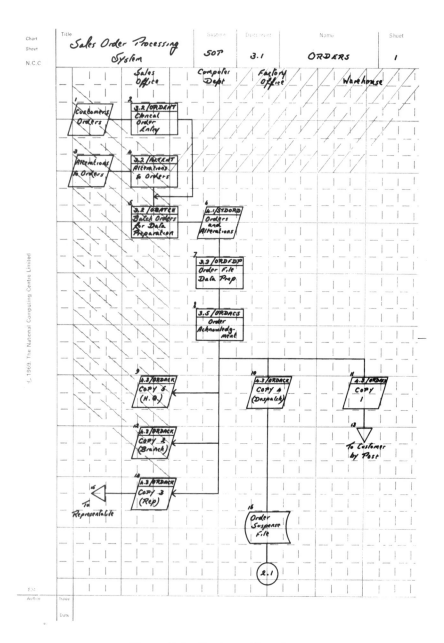

Figure 7.4   Systems Flowchart

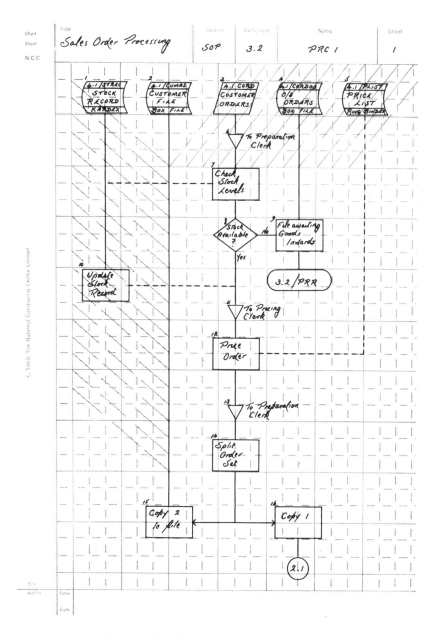

Figure 7.5   Clerical Procedure Flowchart

The broken lines (fig. 7.5) indicate relationships between operations or entities in different columns without the concept of a sequence which is inherent in the firm line.

Other conventions of flowcharting are:

- crossed lines do not imply any logical relationship;

- where two incoming lines join one outgoing line, they should not join at the same point;

- annotation should be included within the symbol for 'process', 'decision', 'file' and 'input/output', giving an indication of the action, decision, file or document used. The annotation for movement (eg to Pricing Clerk as symbol 11 on fig. 7.5) should appear to the right of the symbol. The terminator symbol should always carry words within it: START, EXIT, or PROCEDURE REFERENCE (if the connection is with another procedure). The connector symbol should always include the number of sheet and symbol connected;

- identification of symbols on a flowchart is aided by numbering them (at the top of the symbol) sequentially from 1 on each page thus each symbol can be referenced uniquely (symbol 5 on page 2 = 2.5). Connectors and terminators do not need to be numbered.

DECISION TABLES

Where more complex interactions have to be portrayed, a decision table often provides a suitable alternative. Decision tables must show all possible conditions and combinations of conditions and the actions to be taken for each. Decision tables are divided into four parts:

Condition Stub	Condition Entry	CONDITION SECTION
Action Stub	Action Entry	ACTION SECTION

The condition stub is a list of all the conditions to be taken into account; condition entries are a tabular representation of the combination of conditions to be satisfied. The Action stub is a list of all possible actions which are taken and the action entries are a tabular representation of the sequence of actions to be taken when particular conditions are satisfied. Figure 7.6 shows a typical decision table. The *row* of numbers at the top of the condition entries serve to identify the different *rules* that are used in the table. The last rule is an example of an ELSE rule, which allows the systems analyst to specify a course of action to be taken if none of the preceding conditions is met.

The three asterisked lines are examples of a useful concept called *extended entry*. This allows the analyst to use more complicated instructions than is possible with the normal *limited entry* format, which only allows Yes, No or Immaterial (–) conditions to be used in the condition entries and X (for the stated action) and – (for 'does not apply') to be used in the action entries. For instance, in Rule 3 the action 'set error code = 1' is performed, whereas in Rules 5 and 6 the action 'set error code = 2' is performed, but only one extended entry action stub is required to specify both these actions. Tables which contain both limited and extended entry rows are called *mixed entry* tables. Linkage to other decision tables is achieved by statements such as 'Go to error routine', as is shown at the foot of fig. 7.6

It is normal practice when drawing a decision table to include remarks or a key (in the case of an extended or mixed-entry decision table) at the bottom of the table, and to indicate the number of conditions (C), actions (A) and rules (R) at the top.

The steps to be followed in constructing a decision table are as follows:

– identify conditions and write down in any order;

– identify actions and write down in *correct* order;

– identify all *probable* combinations of conditions and create a rule for each combination showing appropriate actions;

– if all *possible* combinations have not been specified, add on Else rule, with appropriate actions (N.B. no condition entries);

– check for completeness by firstly calculating the maximum number of rules required (for a limited entry table this is 2^c where c is the number of conditions; whereas for extended entry tables, $R = N_1 \times N_2 \times N_3$ ... where R = number of rules, and N_1 = number of possible values of condition 1, N_2 = number of possible values of condition 2, etc), and then comparing the maximum with the actual number of rules in the table. NB: if the ELSE rule is present, it must be given a value and, where dashes are present in condition entry rules, these are complex rules (as opposed to simple rules) and the number of rules = $2 \times R$ where R = number of dashes;

– check for ambiguity by looking for two rules with same conditions leading to different actions (especially watching for 'immaterial' which can conceal this);

Stock File Update | SOP | 3.5 | SUDATE | 1

C = 5
A = 13
R = 14

	1	2	3	4	5	6	7	8	9	10	11	12	13	14	
End of Transaction File	Y	Y	N	N	N	N	N	N	N	N	N	N	N	N	E
End of Master Input File	Y	N	N	N	N	N	N	N	N	N	N	N	N	Y	L
Key Comparison T:MO	-	-	<	<	<	<	=	=	=	=	=	>	>	-	S
T code =	-	-	I	I	A	D	I	I	A	D	D	-	-	-	E
IND set	-	-	Y	N	-	-	Y	N	-	Y	N	Y	N	-	
Perform Update Routine	-	-	-	-	-	-	-	X	-	-	-	-	-	-	
Write MO to File	X	X	-	-	-	-	-	-	-	-	-	X	X	-	
Read new MI	-	X	-	-	-	-	-	-	-	X	-	X	-	-	
Move MI to MO	-	X	-	-	-	-	-	-	-	X	X	X	-	-	
Move T to MO	-	-	X	-	-	-	-	-	-	-	-	-	-	-	
Set IND	-	-	X	-	-	-	-	-	-	-	-	-	-	-	
Unset IND	-	-	-	-	-	-	-	-	X	-	X	-	-	-	
Read new T	-	-	-	X	-	-	-	-	X	X	X	-	-	-	
Set Error Code =	-	-	1	-	2	2	3	4	-	-	-	-	-	-	
Go to Error routine	-	-	X	-	X	X	X	X	-	-	-	-	-	-	
Go to End routine	X	-	-	-	-	-	-	-	-	-	-	-	-	X	
Go to Basic Update	-	X	-	X	-	-	-	-	X	X	X	X	X	-	
Go to Secondary Update	-	-	-	-	-	-	-	-	-	-	-	-	X	-	

Key: MI - record in master file input area
      MO - record in master file output area
      T - record in transaction file input area
      IND - indicator that MO holds previously inserted record
      A - amendment transaction
      D - deletion transaction
      I - input transaction

Figure 7.6   Decision Table

	Title	System	Document	Name	Sheet
System Outline	Sales Order Processing	SOP	3.1	SOPSYS	1
NCC					

Inputs	Cross ref.	Processes	Cross ref.
Customer Order Details	4.1/ORD	Clerical Order Entry	3.2/ORDENT
Alteration to Order Details	4.1/ORDACT	Clerical Alteration to Orders	3.2/ALTENT
Order Input File	4.2/ORDIN	Order File Data Prep	3.3/ORDFDP
Order Controls		Computer Order Acknowledgment	3.5/ORDALS
Despatch Detail Input File	4.2/DESP →	Clerical Sales Order Despatch	
		Computer Despatch	

Files			
Customer Index	4.4/CUSIND →		
Product Catalogue	4.4/PROCAT	Outputs	
O/s Orders	4.4/OSORD	Order Acknowledgment	4.3/ORDAK
Doubtful Credit List		Advice/Consignment Note	4.3/ADV
Delivery Cost List		Balance Order	4.4/BAL
Factory Stock List		Invoice Details File	4.4/INV
Factory Order Ledger	4.1/OLED		

Notes.

S 31

Author	Issue
	Date

© 1976. The National Computing Centre Limited

Figure 7.7 Systems Outline

Clerical Document Specification	Document description Factory Order Ledger		System SOP	Document 4.1	Name OLED		Sheet 1
NCC	Stationery ref. SCL 10		Size A4	Number of parts 1	Method of preparation Handwritten		
	Filing sequence Page no.		Medium Loose-leaf ring binder		Prepared/maintained by Factory office clerks		
	Frequency of preparation/update Daily / as required		Retention period 6 months after despatch		Location All branch offices		

Weekly Receipts

	Minimum	Maximum	Av/Abs	Growth rate/ fluctuations
VOLUME 1	314	555	375	Long-term trend unknown
2	275	484	336	Seasonal variations with orders
3	145	259	178	
Total	734	1298	889	

Users/recipients	Purpose	Frequency of use
Factory office clerk (FOC)	To record orders received	daily
	To record completed despatches	daily
	To check that orders are despatched on time	weekly

Ref	Item	Picture	Occurrence	Value range	Source of data
1	Page (number)	9(5)	1 per page	10,000 - 99,999	FOC
2	Date (order received at factory)	99 AAA 99	1 per order 41 per page	valid date	FOC
3	SCL order number (SALKON)	9(6)	as ref 2	100,000 - 999,999	4.1/ORDES
4	Customer Name		as ref 2		4.1/ORDES
5	Due (date)	99 AAA 99	as ref 2	valid date	4.1/ORDES
6	Actual (date despatch completed)	99 AAA 99	as ref 2	valid date	4.1/ADCONS (copy 5)

Notes Ref. 4 : Customer name is usually first line of delivery or invoice name and address, whichever is more meaningful to factory office clerk.

S 41

Author	Issue	
	Date	

Figure 7.8   Clerical Document Specification

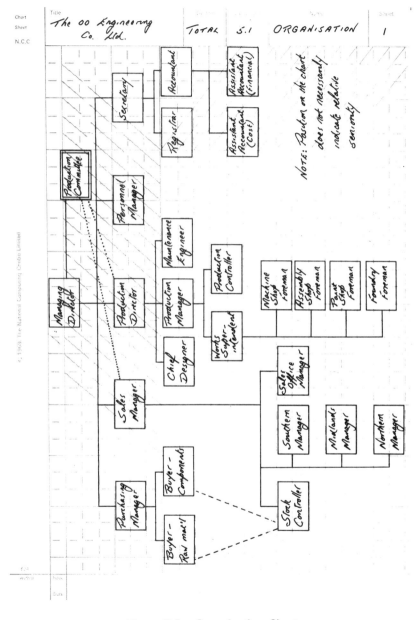

Figure 7.9 Organisation Chart

– check for redundancy by looking for two rules with different conditions leading to same action (here the 'immaterial' entry can be used sometimes to combine them).

Both decision tables and flowcharts can be drawn on the same NCC standard form, the Chart Sheet.

OVERALL SYSTEM

Frequently it is necessary, in collecting facts about a system, to present a comprehensive view of all inputs, files, processes and outputs, independent of sequence of operations, for easy understanding by data processing staff and users. A form which can be used for doing this is the System Outline (fig. 7.7). It can be used both as an aid in checking for duplication and discrepancies and as an index for more detailed documentation through the use of cross-reference columns. Clearly the systems flowchart must be consistent with the system outline.

DATA

The main source of information about data is the file of clerical documents. Each document needs to be identified and analysed by the systems analyst with the assistance of the user department. The analyst obviously needs to obtain copies of each document, and also requires other information about its usage. The Clerical Document Specification (fig. 7.8), provides a checklist of factors which the systems analyst should consider.

The specification should always be completed and filed with the actual document which it specifies. The main points which need to be recorded are the contents, organisation, size and usage patterns: each of these is covered in the Clerical Document Specification.

Occasionally the systems analyst will be investigating an existing computer-based system and in this case the data involved will be more complex: there will be a requirement to ensure that documentation of input, files, records and outputs is accurate. Suitable standard forms for recording this information are described later.

RELATIONSHIPS

In addition to procedures and data, the systems analyst needs to record information about relationships between various entities. The main types of documentation to support this activity are:

– organisation charts (which show formal relationships between people);

– physical layout charts (which show the flow of data between people within an office);

– grid charts (which can be used to show any type of relationship).

Organisation charts

A typical organisation chart is shown (fig. 7.9). This is used to depict the reporting structure within a given work unit or lateral relationships between departments whose work is interdependent. A rectangular box is used for individual job titles; a rectangular box with double lines is used for committees. Firm lines are used to indicate reporting structure; dashed lines to show lateral relationships; and dotted lines to show committee membership.

Two important observations need to be made about organisation charts. Firstly, they can never be definitive, as changes occur so frequently. It is wise to put a date on any chart. Secondly, they tend to be contentious documents, and so a rider should always be included, eg 'Note: position in the chart does not indicate relative seniority'. It is rare for the formal structure shown on the organisation chart to equate with the power structure of the organisation.

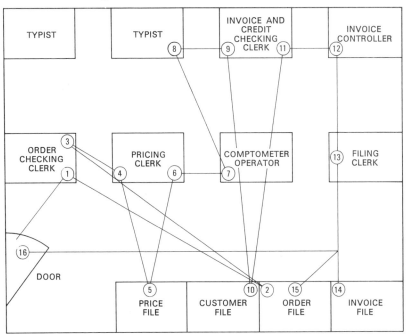

Figure 7.10 Flow Diagram

Physical layout charts

In aiming to improve clerical procedures rather than to computerise them it will often be useful for the systems analyst to draw physical layout charts showing the flow and volume of work. Layout charts can be drawn on the NCC Standard Chart Sheet.

The two main types of layout chart are the Flow Diagram and the String Diagram. The Flow Diagram (fig. 7.10) is used for locating the movements and delays between work stations. The movements are drawn with lines on the physical layout and each work station is numbered in sequence. The diagram shows only one cycle and does not show volumes.

The String Diagram (fig. 7.11) is used to show volumes of work in relation to work flow. It is called a string diagram because the concept behind it is to use string or cotton to represent each time a document passes from one work station to another in the course of a normal cycle; the number of strands represent the work volume.

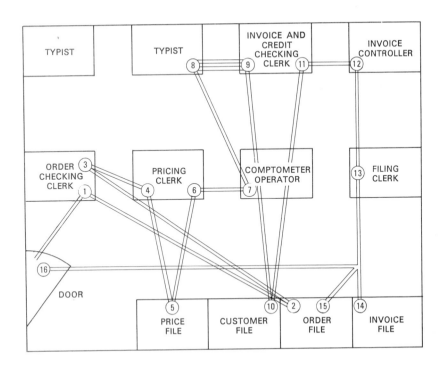

Figure 7.11   String Diagram

The next stage is to evaluate the distance between work stations and the document volumes. These charts are used primarily to assist in the analysis and redesign of office layouts, but they can provide considerable information about the reasons for delays and errors in a system. They pinpoint quite clearly bottlenecks in work flow.

Grid charts

Grid charts are used to show cross- or complex relationships between different sets of factors in a system.

One example is the Document/department grid chart which relates documents used in a system with the departments which either prepare or receive the documents, (fig. 7.12). The intersecting squares are used to indicate the sequence in which each document goes from one department to another. Thus, the customer purchase order goes first to the product manager and next to the sales order office. The sales order is a two-part set. The two parts travel together from the sales order office to typing and back to sales order office where they are split and then go different ways. It will be

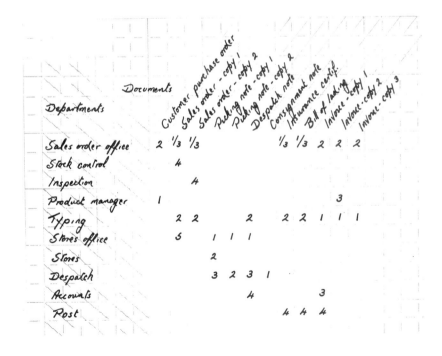

Figure 7.12 Document/Department Grid

clear that this is no more than a rearrangement of information gathered, but in a very concentrated form, and in a presentation which causes questions to spring to mind. This is the essence of good fact-recording.

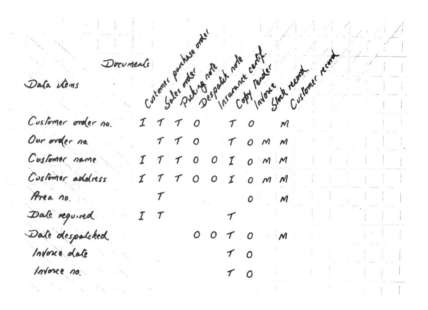

Figure 7.13   Document/Data Item Grid

The same can be said of the Document/data item grid chart (fig. 7.13). This shows duplication of information on various documents, and could indicate possible amalgamation of several documents. This grid uses a different code 'I' standing for input, 'O' for output, 'T' for transfer, and 'M' for master. This may cause additional questions to spring to mind; particularly if an item is input but appears to have no subsequent use, is output but not apparently input, and so on.

Grid charts also can be drawn on the NCC standard form, the Chart Sheet.

SUMMARY

There is a little point in gathering information about existing systems and user requirements unless the findings are adequately documented – both as a record of the investigation which can be agreed with the user and as a basis for subsequent analysis and design work. The recording of the investigation should be based on standard documentation. This should provide facilities for recording information both of an unstructured kind (eg reports on meetings) and in a structured form (eg about procedures, data and relationships). Procedures can be recorded in the form of narrative, flowchart or decision table. Data can be defined on a Clerical Document Specification (also used for clerical files). Relationships between entities can be shown using grid charts or, in the particular case of organisational relationships, using organisation charts. Rigorous documentation of a system aids considerably the analysis of user requirements.

Part III
Logical
System Design

One of the aims of the system investigation is to try to establish the requirements of the users which the new system must satisfy. These are known as the logical requirements, and the systems analyst has to produce a definition of a logical system which will meet the requirements. It is not appropriate at this stage to define the system in physical terms (ie, as a set of computer procedures, computer files, computer devices, office equipment or people); what is needed is a definition of what the user wants the system to do. Once this is agreed, the logical system can be turned into a physical system by more detailed design.

This part, therefore, concentrates on techniques of arriving at a User System Specification. Chapter 8 suggests a methodology for analysing the results of the investigation into a set of requirements; and Chapter 9 describes methods of defining outputs, inputs, stored data and procedures which are appropriate to meet these requirements.

8 Analysing user requirements

INTRODUCTION

Analysis, carried out at an almost intuitive level by the experienced system analyst, is a continuing activity at all stages of the project. It cannot be said to have any true beginning or end other than the beginning and end of the project. As aspects relating to the system and its objectives come to light, they need to be analysed to determine other relevant facts. Equally, as each stage of design is accomplished, the results need to be analysed to confirm that they satisfy the specified requirements and are mutually compatible. There is a main stage of analysis which must occur before a comprehensive specification of requirements can be produced. This cannot be completed until all the relevant facts have been ascertained.

If the same person is to carry out both the analysis of user requirements and the design of the new system, there may be difficulty in reserving judgement on the shape of the new system. Assumed constraints can inhibit the creative questioning which is needed during the main analysis stage.

Analysis may be defined as *the process of dividing into parts, identifying each part and establishing relationships between the parts*. In the specific context of user systems, analysis comprises:

taking known facts concerning a system, breaking these into their elements and establishing logical relationships between the elements, with the objective of producing a specification of requirements.

This can only be done in a disciplined way, using appropriate tools.

During fact-finding, the use of standard forms will help to ensure that nothing conflicts or is omitted. This will enable the systems analyst to modify the planned fact-finding while it is still in progress. This is the first stage of analysis: to examine the data collected and to check that it is complete and correct. This ensures that the data on which conclusions will be based is sound. Part of this phase is to allow the user to examine a picture of the existing system, and to clear up any misconceptions. Once the user has agreed on the statement of the existing system, analysis proper can commence.

The approach must be a questioning one, as each part of the existing system and the proposed new system is examined. The aim is not to find minute flaws but to devise a system which will provide the management and operating information required by objectives and policies, and to identify those parts of the system which can be improved. Change for its own sake should be avoided, but what is taken for granted should be scrutinised.

A useful aide-memoire for this questioning approach is the 'question box':

Primary questions		Secondary questions	
What?	Why?	What else?	Optimum
How?	Why?	How else?	Optimum
When?	Why?	When else?	Optimum
Where?	Why?	Where else?	Optimum
Who?	Why?	Who else?	Optimum

In applying these questions to each event, item of data, relationship or sequence the systems analyst is forced to be critical in a constructive way.

TOOLS OF ANALYSIS

The tools of analysis are not complex. They consist of lists, structure charts, grid charts and flowcharts.

Lists

Objectives, decisions, data, etc, need to be recorded. If each such item is a separate entity (ie no item is a subclass of any other), a simple list of the items and their attributes is adequate. The items of data on the Clerical Document Specification (fig. 7.8) provide an example.

Where one item in a list can be divided into several items at a lower level of detail, it is necessary to show this in some form of structure.

Structure charts

Probably the best known representation of a structure is the 'family tree', as used in the organisation chart (fig. 7.9). Where the structure is very

Managing director ─── Purchasing manager ─── Buyer, raw materials
                                                 Buyer, components

                       Sales manager

                       Production director ─── Chief designer
                                                 Maintenance engineer
                                                 Production manager ─── Production controller
                                                                          Works superintendent ─── Machine shop foreman
                                                                                                    Assembly shop foreman
                                                                                                    Paint shop foreman
                                                                                                    Foundry foreman

                       Personnel manager

                       Secretary ─── Registrar
                                      Accountant

Figure 8.1 A Structured List

complex, or where there are many items at one level, a structured list provides a more convenient and compact format. Fig. 8.1 is the same organisation shown as a structured list.

Having divided the factor we are considering into its component parts, we now need a means of systematically comparing each part with each other part. The appropriate tool is the grid chart.

Grid charts

Grid charts were examined in the last chapter as an aid to fact-finding. Their use in analysing facts is now indicated.

A grid chart consists essentially of two lists, one entered vertically, the other horizontally, with the intersecting squares showing the relationships. Different sets of facts are entered, and different codes used to show different relationships. Thus the grid chart can assist in moving from one complex set of factors to the next.

Flowcharts

Some examples of flowcharts were given in the previous chapter. A flowchart caters for both parts of the analysis process. It divides a procedure into its parts and shows the logical relationships between the parts. It can also show relationships between steps in a procedure and the data used by, or produced by, the procedure. A flowchart can be analysed in detail, with each symbol being queried ('How does the operation, document, or file contribute to system objectives?').

STEPS IN ANALYSIS

The logical steps in analysis are shown, in outline, in figure 8.2. At each step it is necessary to:

- identify the relevant facts, and establish the relationships between them;
- compare that set of facts with the set at each adjoining step and establish the relationships between the facts in these sets.

Objectives results

Analysis starts with the objectives of the system. Before considering details such as data or procedures, a list of the results of the system needs to be rigorously compared with a list of the stated objectives. If there is no discrepancy, then either there is no point in continuing the investigation or the objectives have not been properly formulated. An example of such a comparison is shown (fig. 8.3).

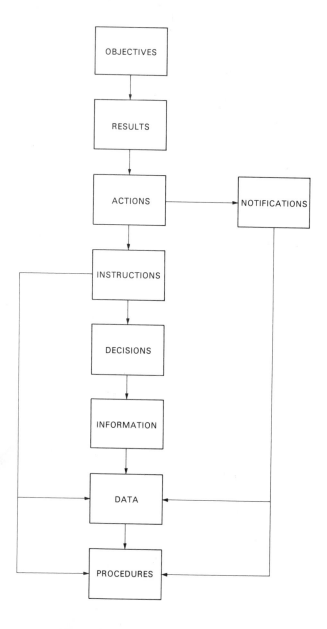

Figure 8.2 Logical Steps in Analysis

SALES ORDER PROCESSING SYSTEM	
OBJECTIVES	RESULTS
1. Fulfil customer orders for spares 80% in 48 hours 95% in 7 days 99% in 10 weeks No order to be outstanding > 6 months	65% in 48 hours 80% in 7 days 94% in 10 weeks 3 orders outstanding > 6 months
2. Provide re-order information within maximum 1 hour of re-order point being reached	30% next day average 2½ hours
3. Provide purchasing department with supplier performance record by Wednesday of following week	Currently 2 weeks late
4. Provide Sales Manager with geographical sales trend graphs not later than following Tuesday	Currently 1 week late

Figure 8.3   Objectives Compared with Results

Results/actions

If the investigation is to continue, similar lists need to be drawn up comparing the results with the actions which cause those results. The actions may be clerical processes, in which case they are within the scope of the systems analyst, or they may be physical processes which fall outside that scope. In either case, if there is reason to suspect that the source of dissatisfaction lies in any of these processes, a case needs to be made to management for an investigation by appropriately qualified staff.

Actions/instructions

Actions other than accidental ones, occur as a result of instructions. The same list of actions now needs to be compared with a list of the instructions which give rise to those actions. The instructions may emanate from a data processing function or from management. They may be ambiguous, indecipherable, mutually contradictory, late or may never arrive at all. If incorrect actions are taken as a result of some fault in the instructions, this needs to be followed up and the faults identified. It is worth noting that there are often several different views of a system given to the systems analyst, eg:

- what the manager thinks is done;

- what the operator thinks the manager wants done;

- what the manager tells the operator to do;

- what the instructions say should be done;

- what the operator thinks the systems analyst wants to know.

Actions/notification

If the data held is to be accurate, there must be a feedback for each action completed into the data processing system (which need not of course be computer-based). The notification may be for individual actions, or for the total number of a given action taking place per hour, day or week. It may be done on paper, by word of mouth, or by some mechanical or electronic means. In order to be sure that this is being carried out accurately, it is necessary to compare the actions taken with a representative sample of the notifications of those actions as received by the data processing system.

Instructions/decisions

If the instructions derive routinely from the data processing system then it is time to start examining the processes. If the instructions come from management, there is a need to compare the instructions which reach the action point with the decisions taken by management, to ensure that there is no failure in communication. Again, two lists should be all that is necessary for rigorous and effective comparisons to be made.

Decisions/information

Decisions are taken (or should be) on the basis of information. Not all the information that a manager uses for a given decision comes from a data processing system. It may come from specialist advisers, external sources, or the manager's own files or memory. It may be influenced by many other factors, such as legal requirements, trade or professional practices, trade union views, the market situation (supply of money, labour or raw materials, competitors' activities or demand for goods and services), as well as by the objectives of the system and the manager's functional and personal objectives. The systems analyst must be aware of all these, but will naturally be most concerned with the kinds of information that the system supplies (or should supply) to management for decision-making.

The same category of information may be of value to several managers, at the same or different times, and in the same or different degrees of detail. Simple lists will not, therefore, provide an effective tool for this purpose. In order to see how these decisions, and this information, inter-relate, we need to construct a Decision/information grid (see fig. 8.4).

Figure 8.4   Discussion/Information Grid Chart

To draw up such a grid, an intimate knowledge of the system and its terminology is required. If the analyst does not have this, then the co-operation of appropriate staff is essential.

The grid is by no means a final answer; until the systems analyst fully understands the reasons for and the significance of the decision, and the exact part played in the decision by each item of information, no precise requirement can be specified. This means spending a lot of time with each manager, unless it is possible to bring a group of managers together in an 'Information Workshop'. In such a workshop, if the purpose and method are properly explained, a lot of the work of analysis will be carried out by the managers themselves: ultimately both they and the analyst will have a real understanding of the uses of information which the system is to produce. This is only likely to come about through a relentless questioning of the precise relevance of each item of information to each decision.

Having determined what information is required, attention then needs to be paid to frequency and urgency. If a given manager requires a number of items of information at weekly intervals, perhaps a single document will suffice. (It is appropriate, at this stage, to secure the co-operation of the manager concerned in producing a draft layout of what the manager would like to see.) If, in contrast, a particular kind of decision has to be made within a few minutes, then either the system must be capable of fast

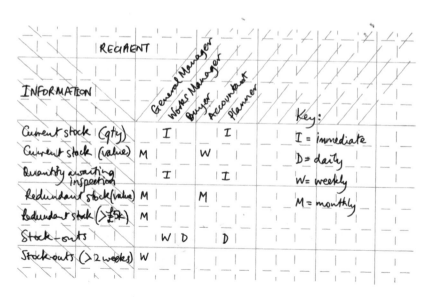

Figure 8.5   Recipient/Information Grid Chart

146 INTRODUCING SYSTEMS ANALYSIS AND DESIGN

response or a visible file of data needs to be constantly available for reference. This kind of analysis is illustrated by the recipient/information grid chart (fig. 8.5).

Information/data and data/procedures

Information is data organised in meaningful form. The various items of information required now need to be compared with the items of data available. The source of such data may be outside the system (from suppliers, customers, government departments, etc) or from within the system, in the form of notifications. The data may be input on each occasion or may be stored as master data within the system.

The formal comparison is carried out using an Input/Output grid chart. A separate grid is drawn up for each individual output (fig. 8.6).

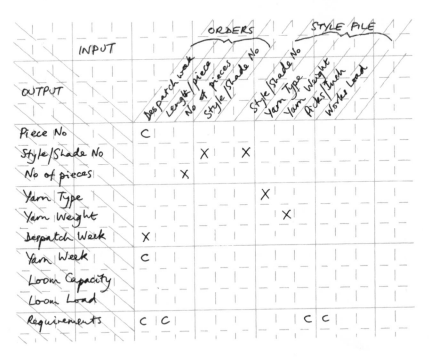

Figure 8.6 Input/Output Grid Chart

Some of the items required as output may be identical to items already in the system, while some may be produced as a result of some procedure acting on one or more items in the system. These different categories will become apparent as a result of the comparison made on the Input/Output grid.

Analysing the procedures

The main check to be done on any procedures developed at this stage is to see that no unnecessary steps are included. The method consists simply of questioning what is achieved by each procedure.

A comparison also needs to be made between the procedures required to produce routine outputs and those required to produce management information, particularly with regard to frequency. If the system is to be computer-based, effort has to be devoted to identifying the procedures suited to computerisation, and those to be handled manually.

Analysing the organisation structure

As part of all the above steps the systems analyst will be considering the appropriateness of the current organisation structure, but at some stage this question must be tackled specifically. Initially the analyst must take a conceptual approach, treating the organisation as though new, and ask the questions 'why does the department exist? why does the application exist? what is its purpose? is it fulfilling that purpose efficiently?' Then the activities of each department must be examined carefully. The analyst must look for unnecessary or redundant functions, duplication of responsibilities (eg between head office and branches), too much or too little communication, complaints from people who deal with the department, high absenteeism, and frequent, heavy overtime. These are all indicative of a need for changes.

Statistical analysis

Statistical techniques should be used whenever applicable, eg to determine sample sizes, to classify data into useful, logical groupings, to isolate the important/significant from the irrelevant, to compare relative values, and to illustrate and compare trends and rates of change. This quantitative information will assist in the design of the computer system as well as pinpoint problem areas at the analysis stage.

SUMMARY

The analysis of user requirements takes place throughout the investigation and outline design stages of a system development project, but at some point in time it should be formally documented. This chapter has suggested a series of steps in analysis using such tools as lists, structure charts, grid charts and flowcharts. These steps start with the system objectives and work through to the data aid procedures. The objectives must be stated in quantifiable terms so that they can be assessed in relation to the results currently achieved; the results are then traced back to the actions required for their achievement, and the actions to the instructions which prompt them and the notifications which confirm that they have been carried out; the instructions must then be analysed in relation to the decisions which produce them, and the decisions in relation to the information on which they are based; finally the information needed can be analysed into the data and procedures required to produce it. This analytical activity will highlight the failings of the current system and the implications of user requirements for any new system. The next stage is to define the requirements in the form of a new logical system.

9   Logical system definition

INTRODUCTION

As a result of analysing user requirements, the systems analyst has some conception of the new system. The next stage is to define the system in terms of a User System Specification (the report is described in detail in Chapter 22). This process is called *logical system definition* because the system is still being conceived in logical terms (ie what the user requires); it has not yet been converted into physical terms (ie how the requirements are to be achieved in terms of hardware, software, equipment, people, procedures, etc).

The systems analyst's technical knowledge will inevitably affect this process. At the same time, the aim is to avoid letting technology dominate or dictate either the identification of requirements or the design of data processing systems to meet these requirements (fig. 9.1).

In logical system definition, a number of factors have to be considered:

- the analyst must clarify the specific objectives of the design process;

- the results of analysis must be converted into an outline of outputs from and inputs to the system;

- it is necessary to structure the data which will need to be stored in order to produce the outputs;

- it is necessary to partition the outline system into computer and clerical subsystems;

- it is necessary to consider the nature of the processing needed to meet the user requirements (eg batch processing *vs* on-line processing).

These requirements are considered in this chapter.

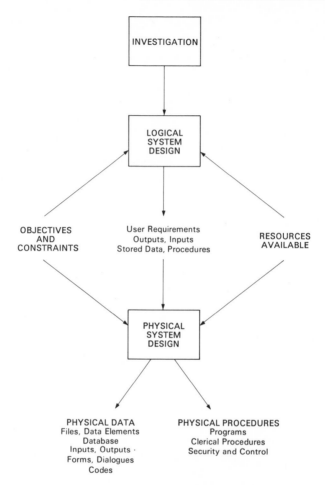

Figure 9.1   Logical and Physical System Design

Design as a creative activity

While analysis is essentially logical, design is essentially creative. The elegant design achieves its objectives with minimum use of resources. Resources are always limited, and the acceptable design often proves to be a compromise between a number of theoretical ideals and resources available. To say that design is a creative activity does not mean that it consists simply of a series of bright ideas. Design requires a full understanding of the problem: there is a need for analysis of requirements and resources.

The essential feature of the creative process is the association of previously unconnected concepts. A technique which may produce results for the individual working in isolation is the one known as 'morphological' analysis. This consists of listing all the elements of the problem and associating each item of the list with each other item. The grid chart provides a useful way of laying out the two lists.

If there are several people involved in the problem area, then the technique of 'brainstorming' may produce useful results. The essential features are:

– no idea is subjected to criticism;

– all ideas are acceptable;

– the wilder the idea the better;

– combinations of ideas and improvements of ideas are sought.

Following a lapse of several days, the ideas are evaluated. Ultimately an idea has to be evaluated against other ideas and checked against the user requirements and the available resources. Finally, it has to be checked against the alternative uses of the same resources. This may seem to reduce the creative element to a small position in the total process; it is still an essential element.

DESIGN OBJECTIVES

At an early stage in defining a new system, the systems analyst must have clear understanding of the objectives which the design is aiming to fulfil. These objectives must be established by management and included in the terms of reference for the project. There is usually more than one way of achieving a desired set of results. The acceptable design is likely to be a compromise between a number of factors; particularly, cost, reliability, accuracy, security, control, integration, expansibility, availability, and acceptability.

Cost

Cost is associated with the two activities of *development* and *operation*. Development comprises all the stages from initial design to successful implementation. Operation includes data preparation, processing, and handling of output and consumables (particularly paper). Operation also includes maintenance of the system: it is very rare for any computer-based system to run for years without substantial changes, to take account of changing business requirements or changing hardware and software. In fact, it is the experience of many long-established computer installations that the major part of their analysis and programming effort is devoted to maintaining existing systems. Clearly the first objective is to minimise costs in these areas.

Reliability

This includes the robustness of the design, availability of alternative computing facilities in the event of breakdown, and the provision of sufficient equipment and staff to handle peak loads (whether seasonal or cyclical).

Accuracy

The level of accuracy needs to be appropriate to the purpose. For instance, the accounts of an individual customer will probably be kept to the nearest penny, whereas the monthly sales for a region may only be required to the nearest £1000. For each defined level a balance needs to be achieved between avoidance of error and the cost of avoiding the errors.

Security

There are many aspects to security, but the ones which particularly concern the systems analyst are confidentiality, privacy, and security of data.

Confidentiality

Some information, vital to the success of a firm, could cause severe damage if it reached the hands of competitors. The system has to ensure that only authorised staff have access to such information.

Privacy

This concerns information about the individual employee or members of the employee's family. If the personnel file or the payroll file is held on the computer, the design of the system must guard against any unauthorised access to the information.

Data security

If the data held on computer files is incorrect, then the system objectives cannot be achieved. Measures are needed to guard against alteration or destruction of data, whether accidental or intentional.

Control

The system should give management the facility to exercise effective control over the activities of the organisation. One way is the provision of relevant and timely information, particularly by extracting the important information from the mass of available but less important information: this is the principle of 'exception reporting', able to produce *ad hoc* reports. Another essential approach is routine control of goods and monies handled.

Integration

Before the advent of a computer, it is usual for each department to keep its own records. This allows the individual departments immediate access, but often causes disagreements. This can result from errors in copying, from source documents being mislaid, or simply from the time-lag between an action occurring (for instance, an item being withdrawn from stock) and the occurrence being entered on the various sets of records.

The ideal, when a computer-based system is being designed, is for all files relating to a given item of information to be automatically updated as a result of a single input. Where this is achieved, it becomes even more important than before to avoid any source document being mislaid before the updating takes place. This means that the systems analyst must be concerned, not simply with the computer procedures, but also with the clerical procedures in user departments and the computer operations department (ie where effective control over the source documents needs to be exercised).

The different systems which make use of any item of information may be designed at different points in time: there is a need for consistent documentation, regardless of who is doing the design.

Expansibility

Any system needs to be able to cope with seasonal or cyclical variations in volume. Estimates need to be made of volume trends, and there has to be provision for handling whatever load the trends predict for the expected life of the system.

Availability

It is the responsibility of the systems analyst to ensure that all the resources required to make the system work are available at the planned implementation time. These can include buildings, hardware, software, stationery, computer operations staff and procedure manuals, as well as the fully tested and working system.

Acceptability

A system which the analyst believes to be perfect is certain to fail unless it has the support of the user departments and management, as well as the support of programming and computer operations.

A system which affects people (and there are very few systems which do not) needs to be agreed by staff representatives, possibly the relevant trade unions, as well as by the individuals whose jobs will be changed. A system which concerns use of assets needs to be acceptable to the auditors.

OUTLINE DESIGN OF OUTPUTS AND INPUTS

During analysis, conclusions should have been reached on the content, frequency and urgency of outputs, both for information required by management and for routine documents, such as pay slips, advice notes, invoices. The analysis should also have shown the relationships between the individual items on each output and the items available as inputs to the system. Where the item to be output is different from the item available as input, the analysis should also have identified the logical steps required to produce the one from the other.

The output reports and input documents should be documented in terms of data content and approximate layout; it is not necessary to define the method of presentation (eg visual display messages or printed lines). It is possible to work back from the output contents, through the system, to the inputs required. This is done by determining which output data items are derived by calculation or by logical deduction; all other items can temporarily be considered to be input to the system. These input items can then be broken down into those which require fresh input every time, as part of input transaction documents, and those which it is more appropriate to store on file because they are historical or relatively static. For example, in designing a sales invoicing system, one might begin by listing the data items required on the invoice, eg:

- invoice number;
- date;
- order number;
- customer number;
- customer address;
- product number;
- product description;
- quantity delivered;   repeated for each
- price;   product line delivered
- value;
- VAT rate;
- VAT amount;

– total value;

– total weight.

If the user agreed with this content, the analyst might then examine the system in terms of what input to the system would cause an invoice to be produced (ie a delivery note) and what data items would be needed on this delivery note. Clearly certain data items on the output invoice are derived (ie VAT amount, value, total value, total weight – all of which are calculated from other data items; in the case of total weight, the calculation is made from an item of data which does not itself appear on the invoice, ie weight per individual product).

The next step is to examine the remaining data and to decide which needs to be freshly input each time and which can be stored on file. In this example, customer address, product description, price, and VAT rate are clearly static data items which could be stored. The customer address would then be identified via the customer code and the product description, price and VAT rate via the product number. In other words the system begins to look like this:

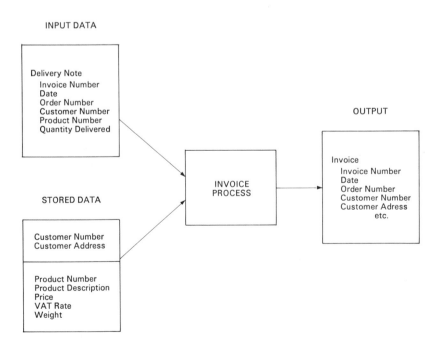

In carrying out this outline design, some of the processes involved (eg calculation of total weight, total value, etc) have also been identified. If another system was under consideration for maintaining the sales ledger, whose inputs were customer payments and sales invoices, then the output information from the example above, as well as being output from the invoicing system, might be input to the sales ledger system. In this case, the invoice details also would need to be stored on file.

DATA DICTIONARY

The identification of data items (as described in the last section) has to be taken a step further in the design of the logical system. Data is the key to the overall system design and must be structured to meet the user requirements.

In specifying and structuring data, the systems analyst is building up a picture of the organisation's data in what is normally called a 'data dictionary'. When all of the organisation's subsystems are computerised, the data dictionary will define the total data resource of the organisation. The data dictionary can be a manual file describing data items, or a computerised system. The document recommended by NCC for defining data items is illustrated in figure 9.2.

The concept behind this form is that each item of data should be uniquely identified and defined; and that it should be cross-referenced to the records in which it appears and the programs/procedures which use it. The complete set of data definitions can provide the basis of the data dictionary. It facilitates cross-referencing and assessment of the implications of changes. Ideally each item of data should be uniquely identified by the data definition, but in individual application systems the same item of data may be used in different ways and under a different name. Therefore, the form allows cross-reference to other names for the item as used in other systems. For groupings of data items into records a record specification form (fig. 10.5) can be used; and for groupings of records into files or databases a file specification form can be used (fig. 10.4). If the data dictionary is held manually in the form of a complete set of data definitions it is usually separated from the individual files of system documentation to provide a central reference file.

Once data items have been defined at the lowest level, then the data structure required by the particular system under consideration can be identified. The terms normally used to define structures are:

- *group items*, combinations of data items for some logical purpose (eg the data item 'customer address' might well be combined with the data item 'customer name' to form a group item 'customer name and address');

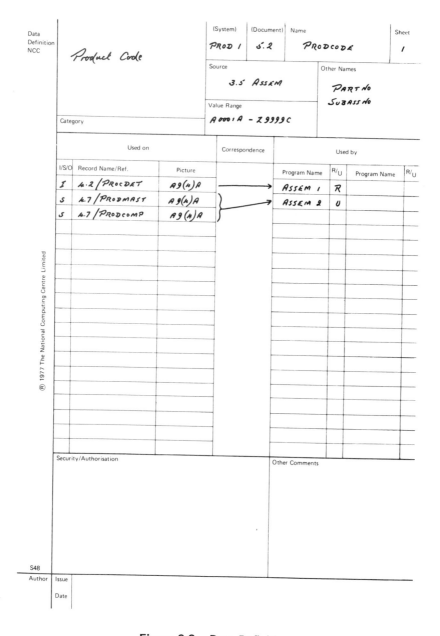

Figure 9.2 Data Definition

 — *records*, collections of data items or group items which are gathered to define an entity for a particular processing purpose (eg the data item 'customer number' and the group item 'customer name and address' might be brought together to form a customer name and address record; whereby in a sales invoicing application the customer number input to the system on a delivery note would be matched with the customer number in the customer name and address record to pick up the name and address);

 — *files*, collections of records which are gathered together to allow processing of data about several entities (eg if in the invoicing system all the deliveries for a particular day were processed to produce invoices then a file of customer name and address records would be required to match against all the different customers to whom a delivery had been made);

 — *database*, a collection of records which are interrelated for purposes of processing to meet user requirements (eg in a deliveries database, the delivery record would be interrelated with the order record, product record, customer record, and invoice record, in such a way as to minimise duplication of data).

In the logical design stage, the systems analyst has to identify the relationships required between the various items of data and the grouping of items of data together into records. This is known as the logical data structure which is required to produce the outputs which users require. It will subsequently be converted into a physical structure (ie medium, organisation and access method) in the detailed physical design. It is important that the analyst does not allow design at this early stage to be over-influenced by the physical format; this may prevent seeing possible ways of meeting user requirements.

Data analysis

One of the most useful methods of analysing the data required by the system for the data dictionary has developed from research into relational databases, particularly the work of E. F. Codd. This method of analysing data is called 'normalisation'. *Un*normalised data is illustrated in the first part of the example in figure 9.3. Here the data describing an employee in a personnel records system is shown in unnormalised form, the training courses entry being a repeating item for each course attended (the key of the record, ie, the identifying data item, is in this case, employee number).

The first stage in normalisation is to reduce the data to its first normal form, by removing repeating items and showing them as separate records but including in them the key field of the original record. Thus in part two

1. **Unnormalised data**

   Personnel record
     Employee Number
     Department Number
     Manager Number
     Training courses attended
       Course Number
       Course Title

2. **First Normal Form**

   Employee record   Training record
     Employee Number   *Employee Number*
     Department Number   Course Number
     Manager Number   Course Title

3. **Second Normal Form**

   Employee record   Training record
     Employee Number   *Employee Number*
     Department Number   *Course Number*
     Manager Number

                                      Courses record
                                        Course Number
                                        Course Title

4. **Third Normal Form** (TNF)

   Employee record   Training record
     Employee Number   *Employee Number*
     Department Number   *Course Number*

   Department record   Course record
     Department Number   *Course number*
     Manager Number   Course title

Figure 9.3 Normalisation of Data (the record keys are underlined)

of the example, in the first normal form there are now two records, employee record and a training record, with the key field employee number appearing in each.

The next stage, of reduction to the second normal form, is to check that all the items in each record are entirely dependent on the key of the record (ie, in the example, on employee number). If a data item is not dependent on the key of the record, but on another data item, then it is removed with its key to form another record. This is done until each record contains data items which are entirely dependent on the key of their record. In part two of the example in figure 9.3, Course Title is dependent on Course Number and not on Employee Number, and so a Course Record has to be created as a separate record (thus giving the second normal form in part three).

The final stage of the analysis, the reduction to third normal form (TNF), involves examining each record to see whether any items are mutually dependent. If there are any, then they are removed to a separate record leaving one of the items behind in the original record and using that as the key in the newly created record. In the example, department number and manager number are mutually dependent because the manager of each department will always be the same. Thus a new record is created for departments, the key being department number (part four of fig. 9.3). The data is now completely normalised.

Once normalisation is achieved, the next task is to identify the relationships between the records which have been defined. This can be done using a data structure chart, which is related to the appropriate data specifications. In the data structure chart, records (or items of data) are represented by a rectangular box incorporating the name of the record (or item); an upper stripe can be used to cross-refer to the relevant record specification or data definition. Relationships are shown by connecting lines; a triangle is used to represent more than one record as in the following examples:

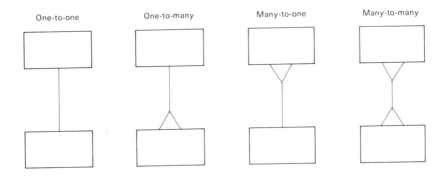

A data structure chart for the example of the TNF records would appear as figure 9.4. Each department record will be related to several employee records (ie, many men in one department); each personnel record could have several training records (ie, each man on several courses), etc. If the data was stored in third normal form then clearly it would lead to a considerable saving of storage, since manager numbers would not be duplicated for each employee in the same department, and course titles would not be duplicated for each employee who has been on the same course. On the other hand, physical relationships would have to be established between the employee record and the department record, (to

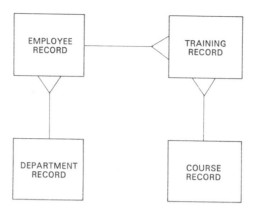

Figure 9.4   Data Structure Chart Example

discover the employee's manager number), between the training record and the course record (to discover the course titles) and between the employee record and the training record (to discover which employees in a department have been on particular courses). Once such relationships are established, it is much easier to process data stored in third normal form than in unnormalised form because several keys to the data are available. If even more keys are required, then other data items in a particular normalised record can be designated secondary keys.

The advantages of carrying out this analysis into third normal form are found mainly in database design where a major aim is to build up related sets of non-redundant data. (This is taken up in greater detail in Chapter 11.) But it is also a useful way of analysing and defining logical data requirements at the logical system design stage.

Normalisation discourages the analyst from viewing the logical data in terms of physical constraints of file sequence, record length, access methods, and yet facilitates the conversion of the logical structure into physical data organisation. Once the data content and relationships have been determined, some estimate must be made of volumes, frequencies, etc, in relation to each record. This is necessary to gain some idea of the scale of the system so that it can be costed and evaluated technically, and also to provide some of the constraints which will be useful in producing the physical design.

Cost of information

Some information is available as a by-product of routine processing; other kinds may require special collection of data or the storage of data beyond that which is required for the routine outputs. Where these involve any substantial additional cost, the user department which has requested the information needs to be informed of the cost of supplying it. They may find that some different, and less costly, information will serve, or that the benefit they expect to derive from the information is of a lower order than the cost of providing it. It is not within the authority of the systems analyst to refuse information which can be provided; equally, it is wrong to provide all the requested information without making the user aware of any special costs involved.

PROCESSING REQUIREMENTS

Once the outputs, inputs and stored data have been defined, it is necessary to decide what procedures are required to process the inputs, keep the records up-to-date and produce the outputs. There are two major areas for concern here: firstly, which procedures should be handled by the computer, and which by human beings; and secondly, what type of computer processing is required.

Partitioning into computer and manual subsystems

This is one of the major problem areas for the systems analyst – deciding which parts of the system can/should be computerised – and is one in which judgement and experience are important. One essential point is that the systems analyst should not set out to 'put everything on the computer'; this is not feasible in most instances, because complex, open systems require human interaction. Nor is it desirable: if the analyst aims to computerise most of the system, the human beings will be left with tasks that are either boring and monotonous (eg data preparation, source document checking) or frustrating for the individuals who cannot control the machine system.

Research has been carried out into the factors which affect job satisfaction and they are many, but a key factor is the nature of the job itself. The analyst must spend time designing jobs which are rewarding to the incumbent. Briefly, this means identifying complete tasks for each individual which present opportunities for good performance, rather than create demands; which place trust in the employee rather than imply inability; which allow decision-making, even at the lowest level, rather than routine drudgery; and which foster learning and improvement rather than stagnation.

The analyst must aim to exploit the best features of both staff and machine. It is important that they act together in an integrated way to achieve the system objectives. The computer must be regarded as a tool to assist human beings to carry out their tasks.

There are certain activities which the computer clearly performs better than human beings. Such activities include storage and retrieval of large masses of data, rapid calculations, decisions involving perfectly accurate and complete data, reliable processing of data, and rapid (and flexible) output. Also the computer is good at following predefined instructions in a non-adaptive way.

Human beings, on the other hand, have complementary skills. They can recognise patterns in events which have not previously occurred and so identify new issues. They are good at making decisions based on experience, judgement and intuition. They have a vast associative memory (though it is not reliable for detailed points). They are very flexible communicators; and they are good at resolving ambiguity and uncertainty. In other words, the human being is self-motivated and adaptive.

The implications of this complementarity between human being and machine are that the human part of the system should handle the exceptions with which the computer cannot deal (such as errors and amendments). It should have a major role in those aspects that require judgement. It should drive the computer part of the system to suit the changing needs of the environment, and it should control the overall system. The greater the control which the user can exercise over the system, the more effective the system will be. The computer part of the system should concentrate on the manipulation of data which the human being cannot handle so well; and it should make available in a straightforward way all the information that the user needs. The systems analyst needs to spend considerable effort in building flexibility into the input and output aspects of the computer system. With the advances in technology, in particular on-line distributed systems, it should be possible to design computer-based systems with which the user, even at the lowest level, can integrate. Most systems currently in operation tend to keep the user away from the computer system, by providing output data for the user to work on independently. The aim must be to achieve teamwork in problem-solving with the computer offering guidance and suggestions to the user.

The components of the computer and manual subsystems are covered in volume 2. The systems analyst's task must be to get the relative contributions of human beings and machine right in the logical design of the system, and then to ensure that the person-machine interface is well designed. The latter involves careful preparation and presentation of codes, documents, dialogues and procedures, as described in Part V – but it also involves well written manuals about the computer part of the system.

Batch or on-line processing

The decision about the nature of processing in the system will be influenced by a large number of factors, including response time, frequency, data volumes, hardware constraints (especially for data collection), security requirements, and cost.

Response time

Response time refers to the maximum acceptable time interval between a request for information and its receipt by a user. There are not many systems which require immediate response after full processing of input data; a day and sometimes even a week is perfectly adequate in most cases. But for those systems which require fast response, some type of on-line system is required.

Frequency

Frequency refers to how often the users require to retrieve information from the computer system. This may be at regular or varying intervals, but is not usually time-critical. For systems which require up-to-date information about the state of a file, again the file will need to be updated in on-line mode.

Data volumes

One of the factors which will greatly affect the cost of an on-line system is the volume of data at specific points in time. For example, if sales orders are entered on-line, then the system must be able to cope with the peak loading which may be seasonal. The order entry load in the weeks before Christmas for a mail order company may require a system for that period which is much larger than is needed for the rest of the year.

Hardware constraints

One of the major constraints on the type of processing is of course hardware availability in the organisation. Data which cannot be easily keyed in (eg that which is suited to OMR/OCR turnaround documents) cannot be handled easily in an on-line transaction processing system. Usually, data collection facilities are the limiting factor in the decision whether to go on-line.

Security requirements

The security requirements of a system can have a significant impact on the nature of processing in the proposed system. Security is more expensive and more difficult to achieve in an on-line system. This makes severe demands on the system designer.

Cost

Finally, cost can be an inhibiting factor. Usually the cost of on-line processing cannot be justified in simple comparison with the previous system. There must be seen to be quantifiable extra benefits which this more expensive method of processing achieves. When the method of processing has been decided, the necessary processes must be identified and broken down into the appropriate clerical and computer procedures and recorded in flowchart form. At this stage in the project, an outline is all that is required.

USER SYSTEM SPECIFICATION

The outline design which has been carried out in defining the logical system will be documented in the form of a User System Specification which is described in detail in Chapter 22. Once this outline has been agreed by the users, it will be turned into a physical system by more detailed design. The activities of detailed design are described in Volume 2. It should be noted that logical system design does not stop at this point; rather it is an iterative activity (fig. 9.5).

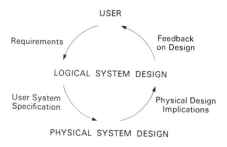

USER

Requirements

Feedback on Design

LOGICAL SYSTEM DESIGN

User System Specification

Physical Design Implications

PHYSICAL SYSTEM DESIGN

Figure 9.5   Interactive Nature of Logical and Physical Design

SUMMARY

The analysis of user requirements will have identified what the new system must be able to do and this has to be defined to form a basis for discussion. The definition of the logical system is sometimes considered to be a design activity (albeit in outline) which requires creative thought, but it is more an extension of the analytical process which has been taking place already. The systems analyst has now to define, in the light of project objectives, the outputs, inputs, stored data and processes of the new system. A key aspect of the new system will be the data structures that are required to be manipulated and so a major task at this stage is to carry out a detailed analysis of data structures. It is also necessary to partition the procedures into manual and computer subsystems and to examine possible approaches to the computer processing. The definition is documented in the form of a User System Specification which is then discussed with the users. Once it has been agreed, detailed physical design of the system can begin.